Upgrading and Operating the KRMx01 CNC

The Illustrated Guide to the Operation of the KRMx01 CNC

Michael Simpson

Kronos Robotics
August 2012

Upgrading and Operating the KRMx01 CNC

KRMx01 Series Book 2

Published by Kronos Robotics

Leesburg, Virgina

978-1-938687-11-2

Dedicated to my wife, Debb.

Forward

In book one of the KRMx01 series you completed the basic construction of the KRMx01 CNC. You wired the KRMx01 electronics and configured the Mach3 controller software. In this book, the second in the series, you will add cable management, an emergency stop switch, clamp table, and homing switches. In addition you will be taken step-by-step through the operation of your KRMx01 CNC.

Table of Contents

!! IMPORTANT !!
Please Read

You will need these files in order to complete some of the chapters in this book. The files can be found here:

http://www.kronosrobotics.com/krmx01/da69781_9386_87112/

In addition you can find much more information on the KRMx01 CNC here:

http://www.kronosrobotics.com/krmx01/

Chapter 1

Getting Started

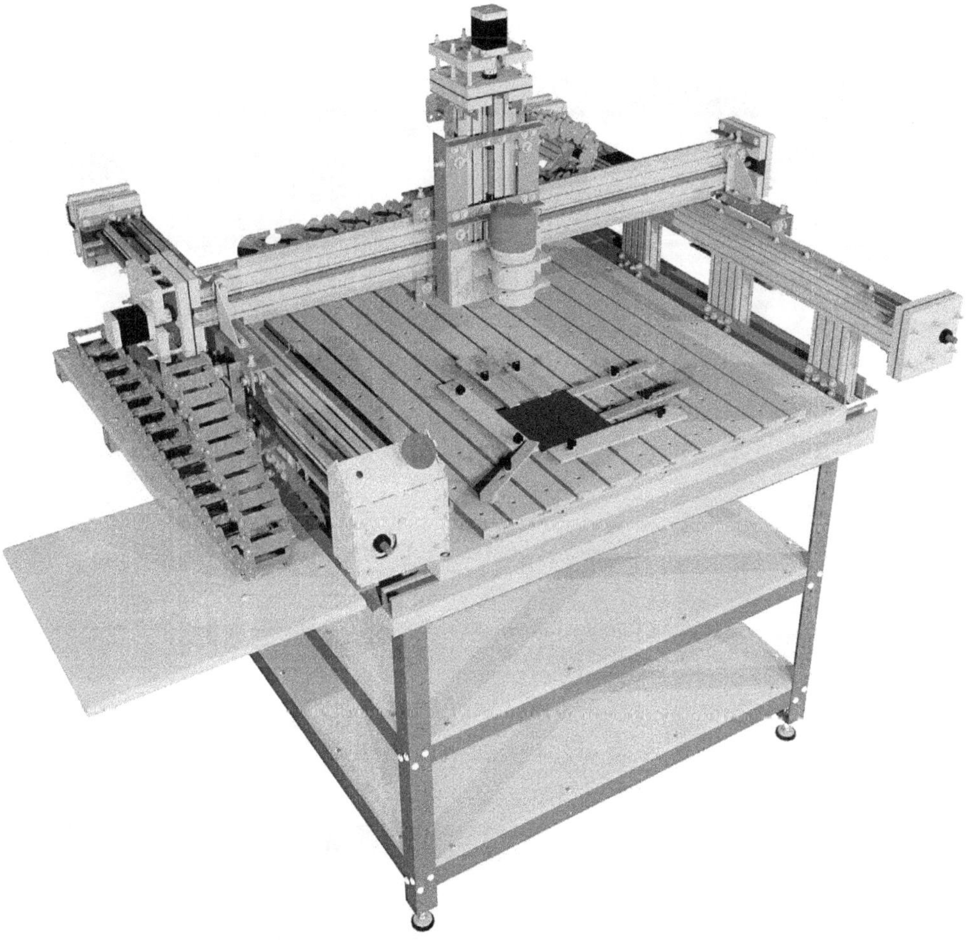

In book one of the KRMx01 series you were taken through the construction process of building the KRMx01 CNC. While the CNC was operational it lacked some fundamental upgrades.

The upgrades presented in this book are crucial building blocks to a well tuned and operating CNC. By adding these upgrades you take your CNC to the next level.

Before proceeding let's take a look at each upgrade in detail.

Clamp Table

The clamp table is essential for securing your stock to the CNC. The clamp table shown in Figure 1.1, utilizes T-slots so that you can easily slide your clamps as needed to securely hold your part while it is being milled.

The clamp table is made from MDF T-slot shelving and is readily available in both full and partial sheets at most home centers.

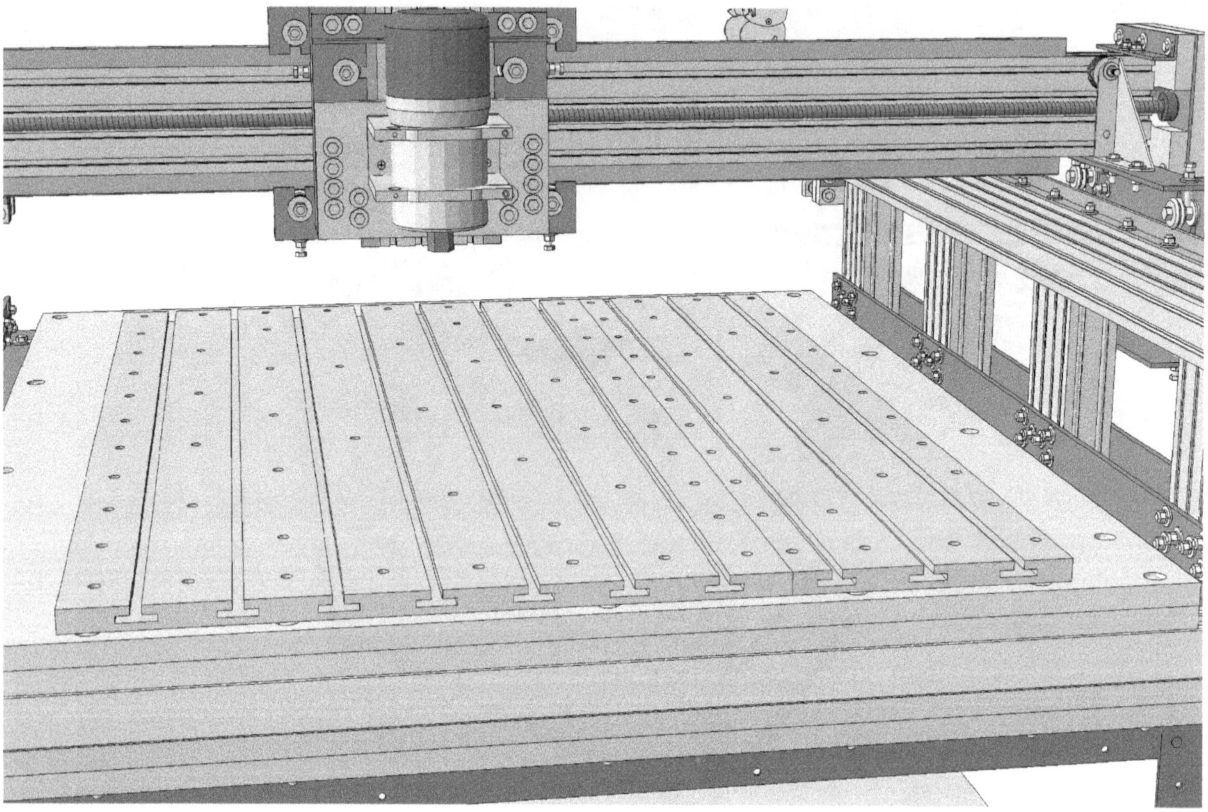

Figure 1.1

Fences

By adding fences to your clamp table, you can quickly reindex your stock. This make it easy to cut multiple parts quickly and accurately. The Fences shown in Figure 1.2, can be adjusted to any location on the table. It is even possible to have multiple fences set up on your clamp table.

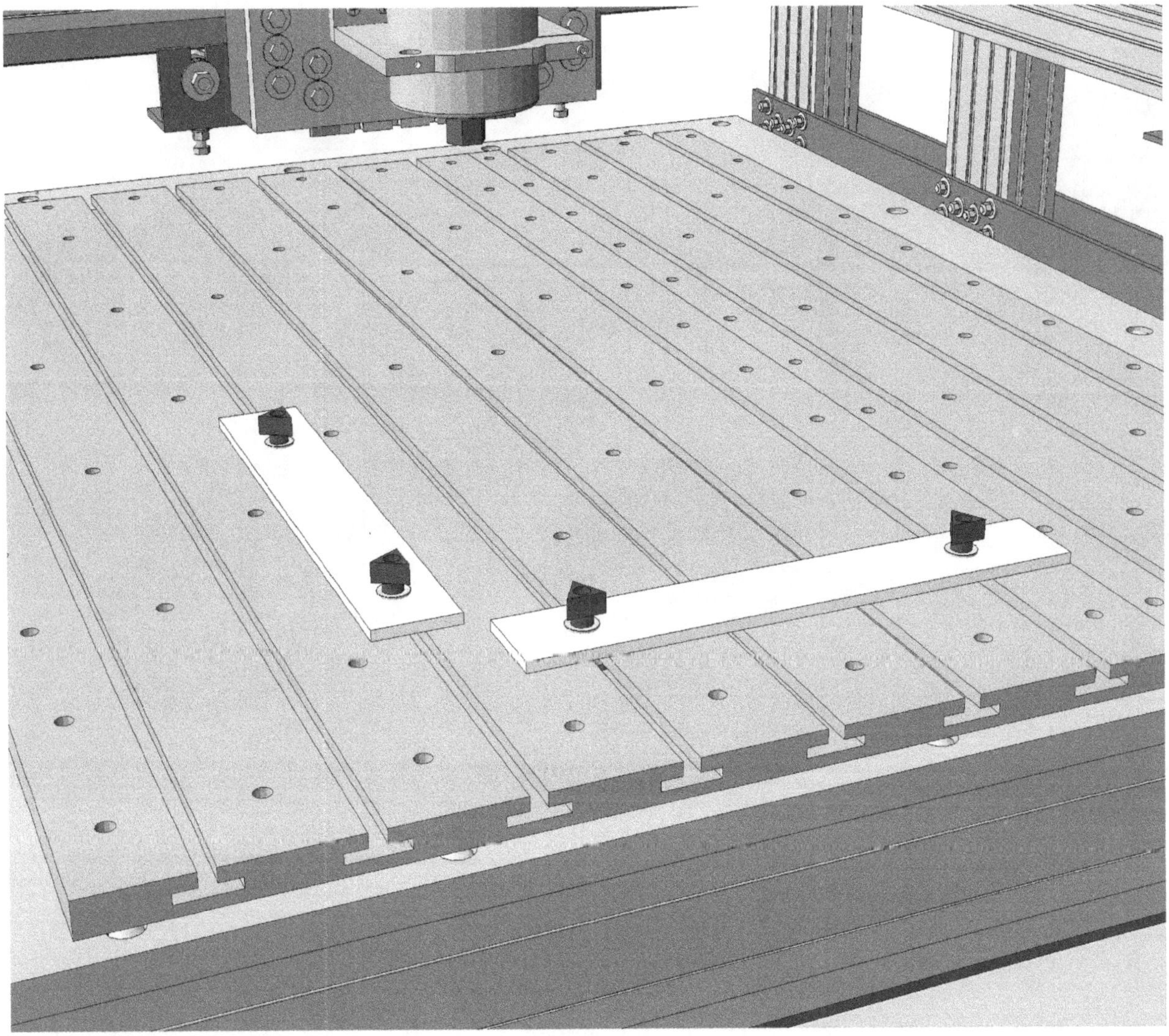

Figure 1.2

Basic Operation and Clamp Bushings

In addition to the upgrades you will learn the basic concepts for creating parts. You will be shown step-by-step how to make the clamp bushings shown in Figure 1.3. The bushings are used to slide into the T-slots in your clamp table. The clamp bushings are used to secure the clamps to the table.

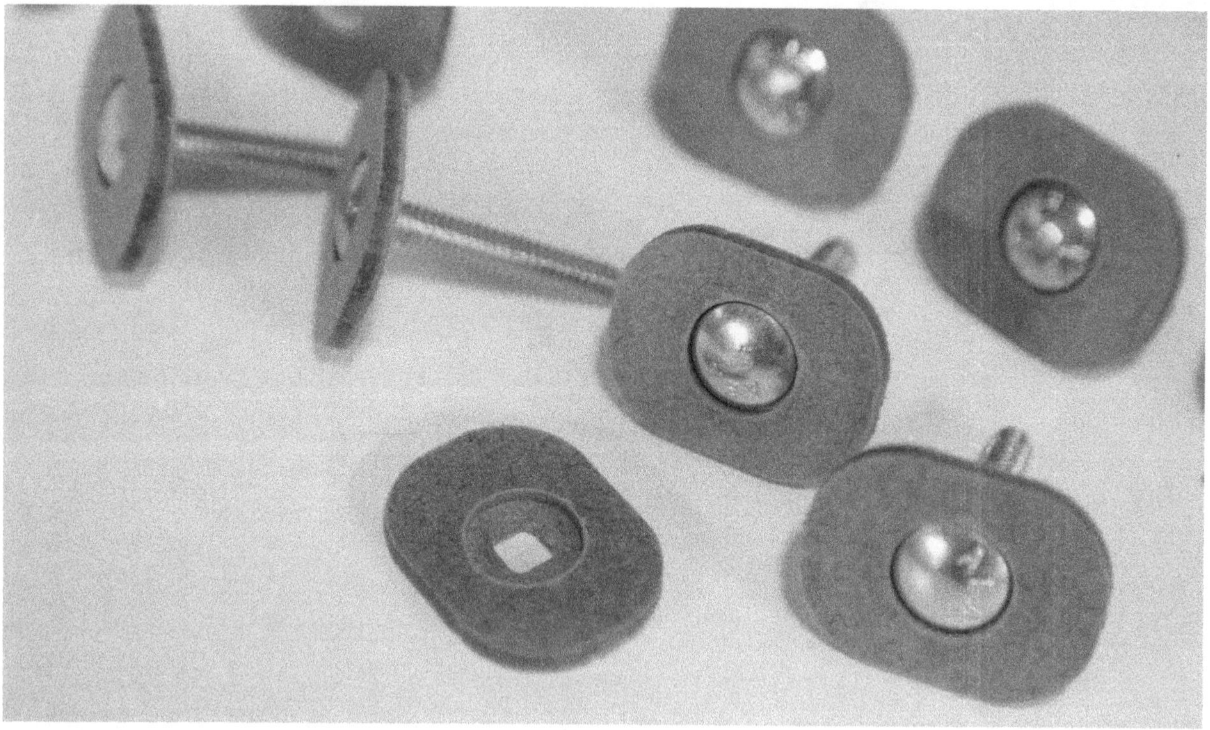

Figure 1.3

Dragon Cable

A good cabling system is essential for the maintenance free operation of your CNC.

The dragon cable shown in Figure 1.4 will enable you to remove your hanging cables. Once attached to your CNC you will run your cables inside the dragon cable. This not only looks better but keeps the cable from getting snagged and cut.

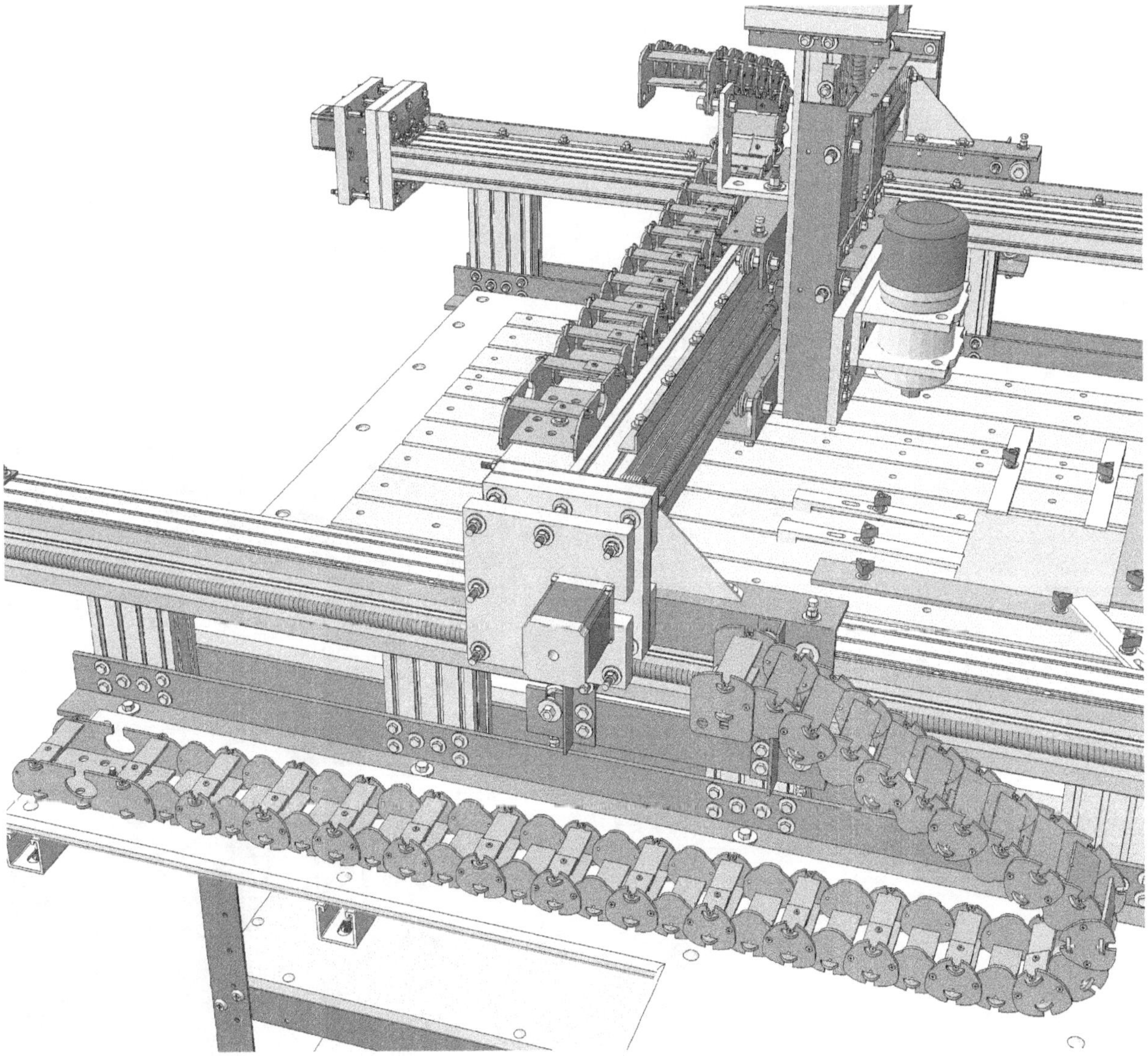

Figure 1.4

E-Stop Box

For safety reasons you need to add an external emergency stop switch to your CNC. This switch will enable you to quickly stop your CNC in case of trouble. The switch shown in Figure 1.5 is located on the front end of your left Y-beam. This puts it within easy reach should you need to hit it in an emergency.

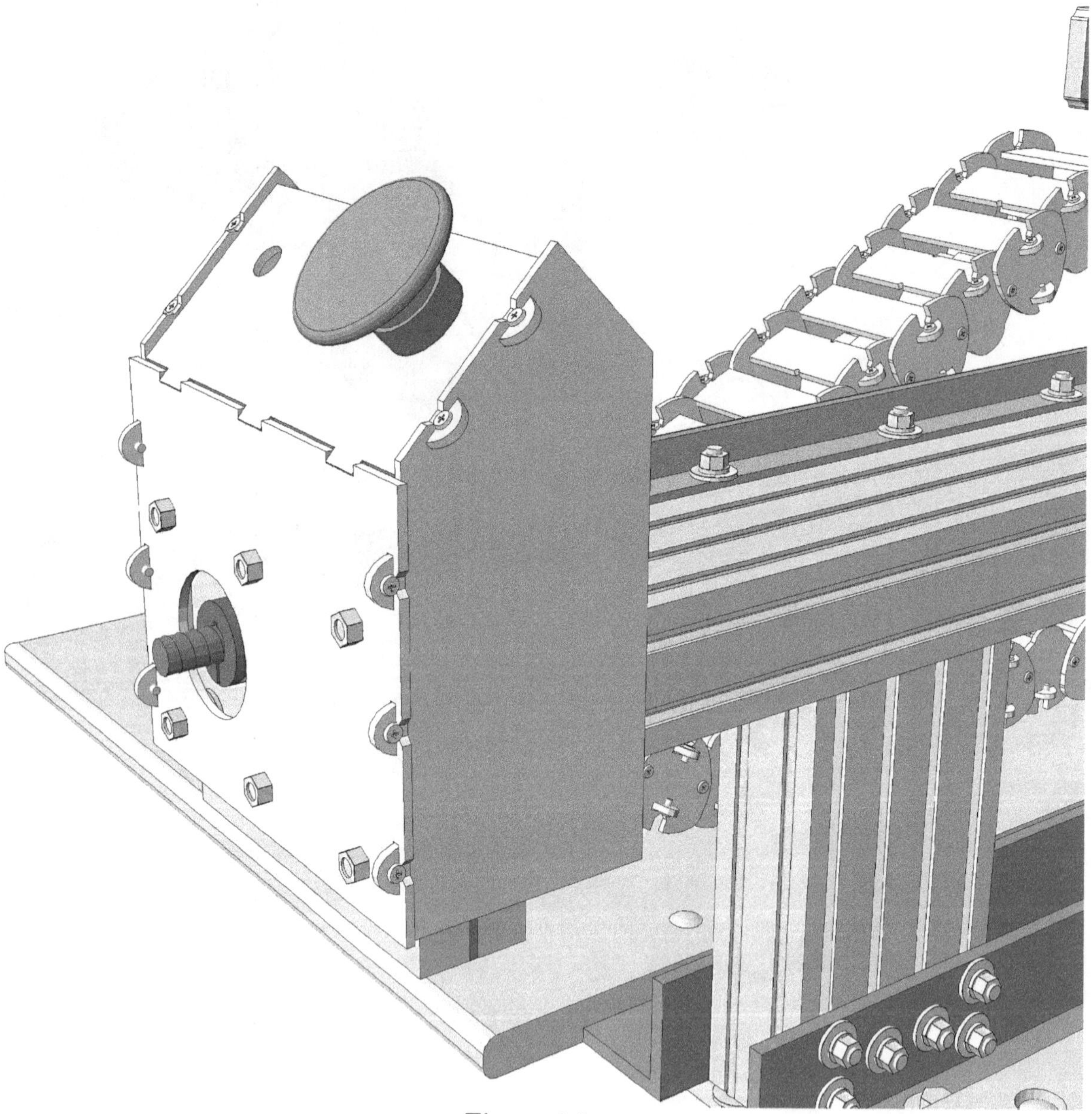

Figure 1.5

Homing Switches

The cornerstone upgrade to the KRMx01 CNC is a set of homing switches. These switches will make referencing your machine as easy as hitting a single button.

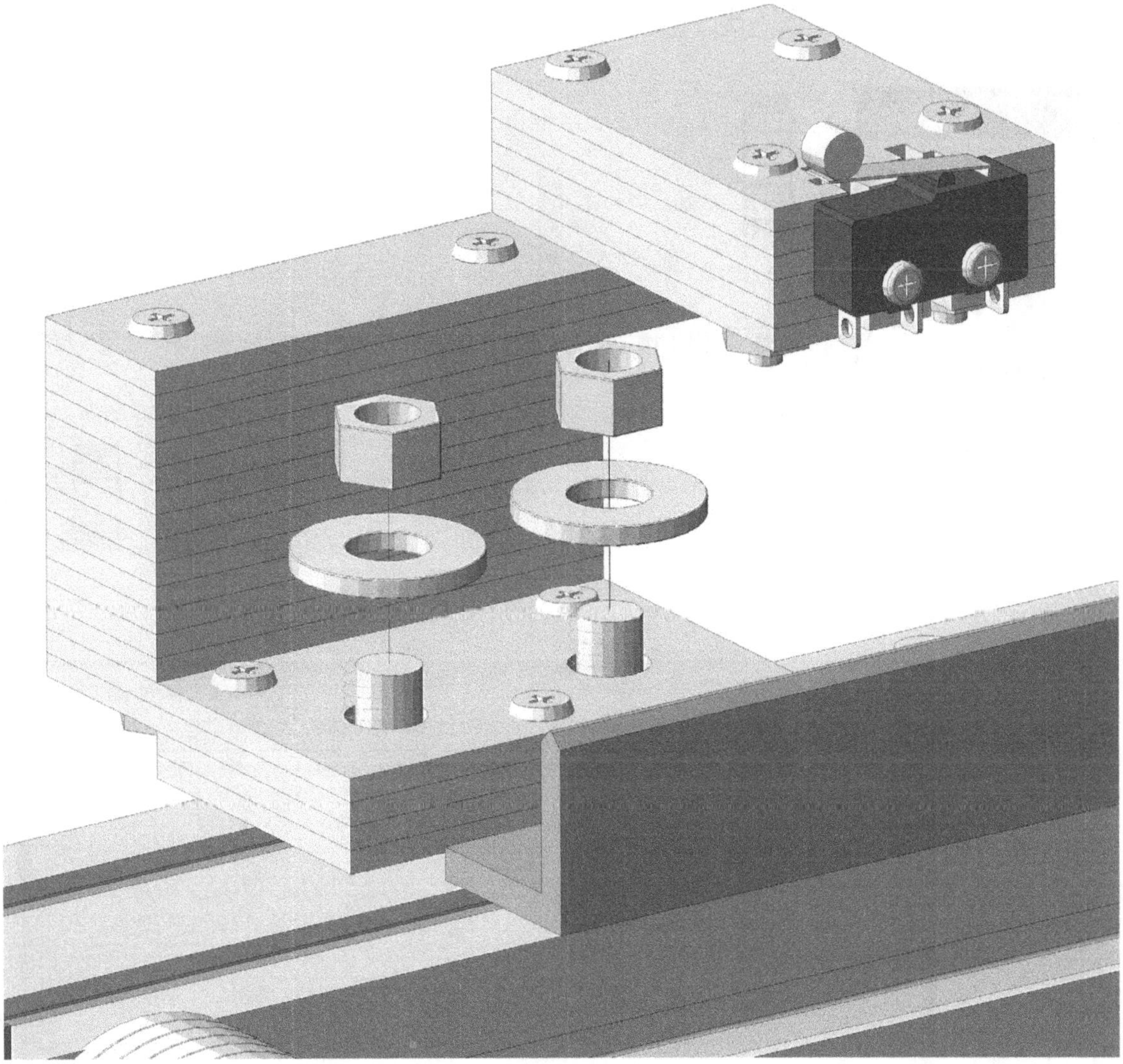

Figure 1.7

The homing switch upgrade also includes extensive wiring directions as well as configuration instructions, as shown in Figures 1.8 and 1.9.

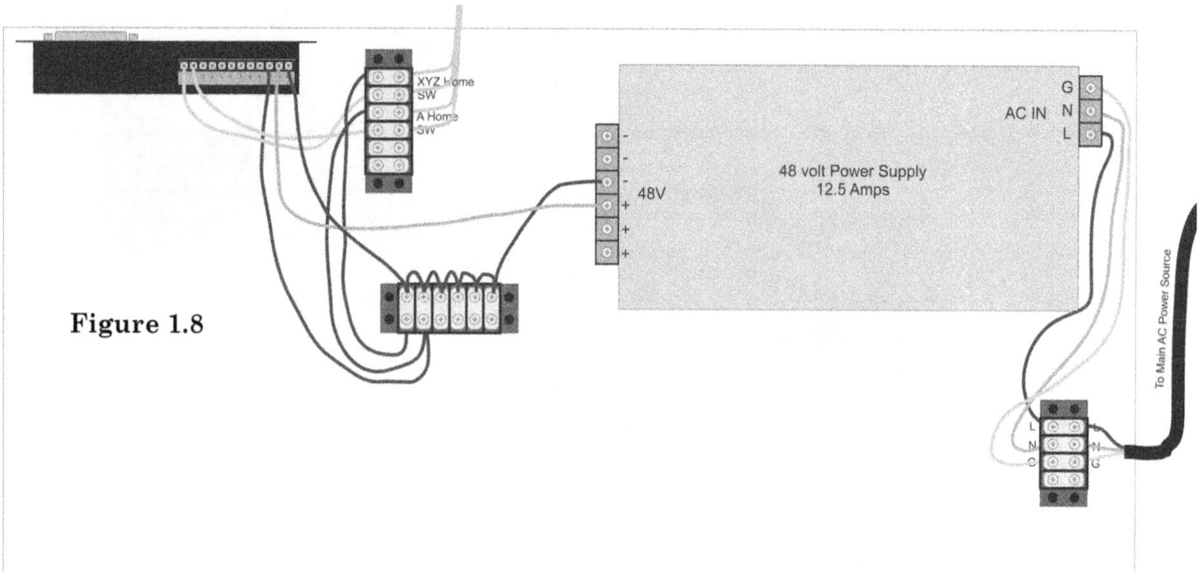

Figure 1.8

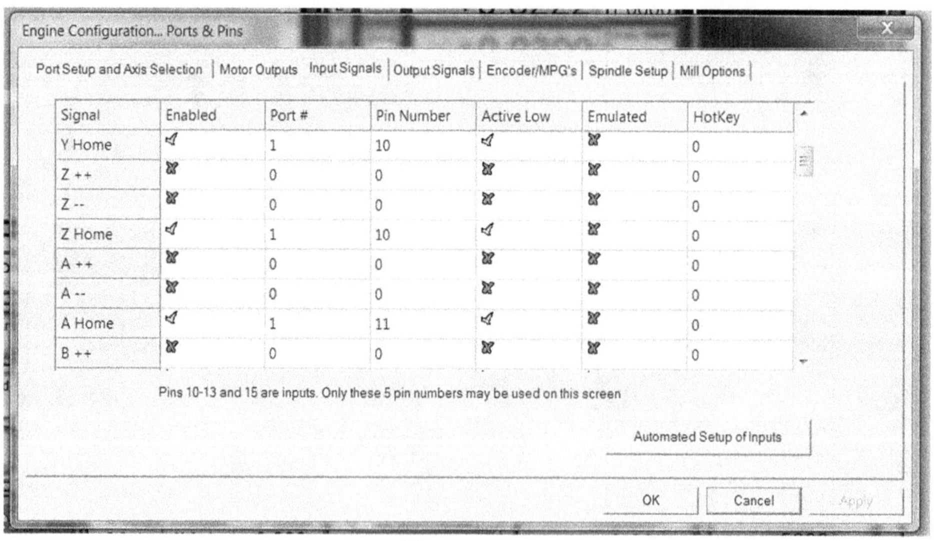

Figure 1.9

Y-carriage Brace

The Y-carriage brace shown in Figure 1.10 is an easy upgrade that will help stiffen your X-beam. This upgrade will allow you to mill faster and more accurately.

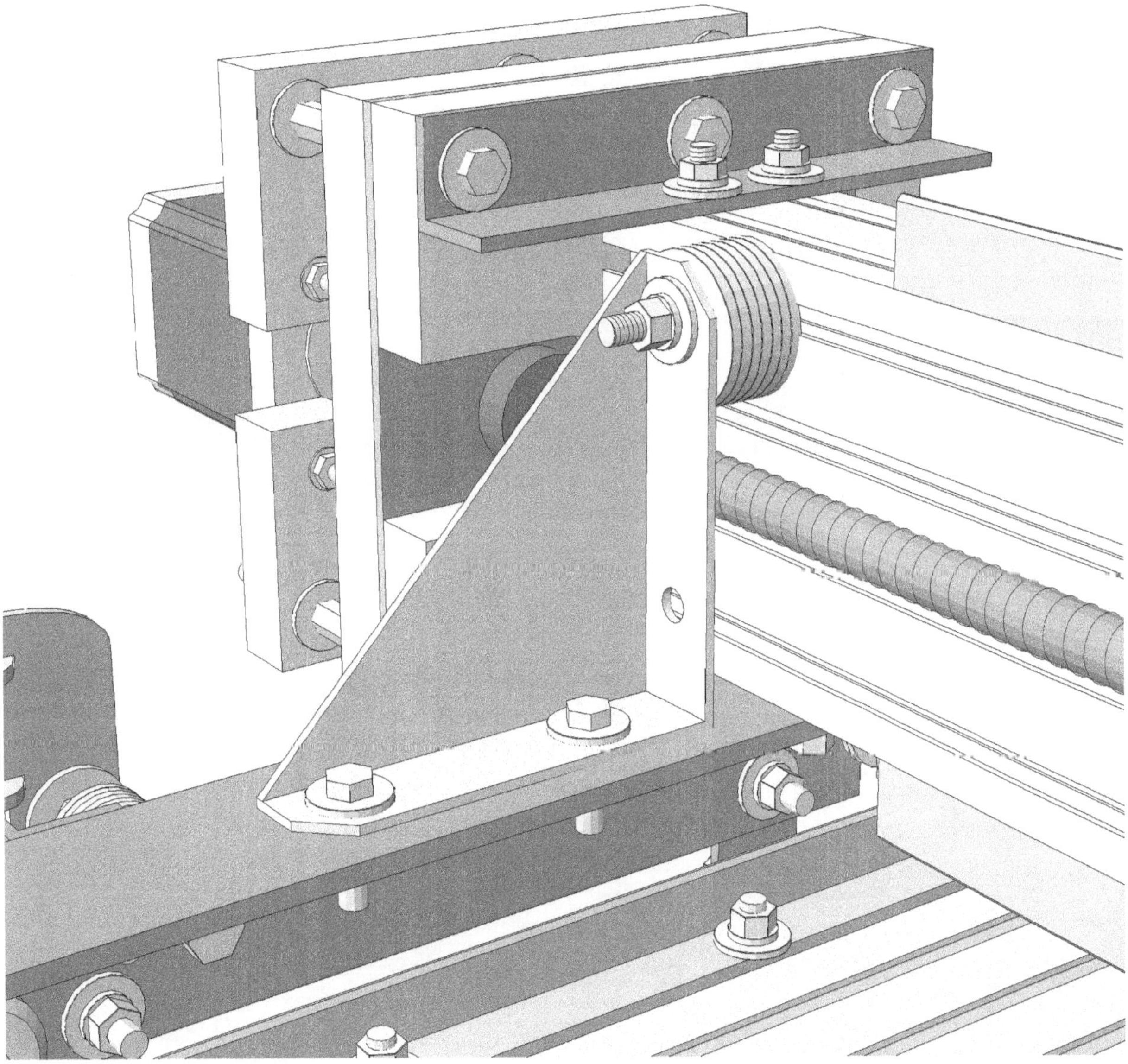

Figure 1.10

Conclusion

Before starting any of the upgrades in this book, visit the books download page.

Upgrading and Operating the KRMx01 CNC download page:

http://www.kronosrobotics.com/krmx01/da69781_9386_87112/

The page has downloads and many shortcut links to components listed in many of the chapters.

Chapter 2

Clamp Table

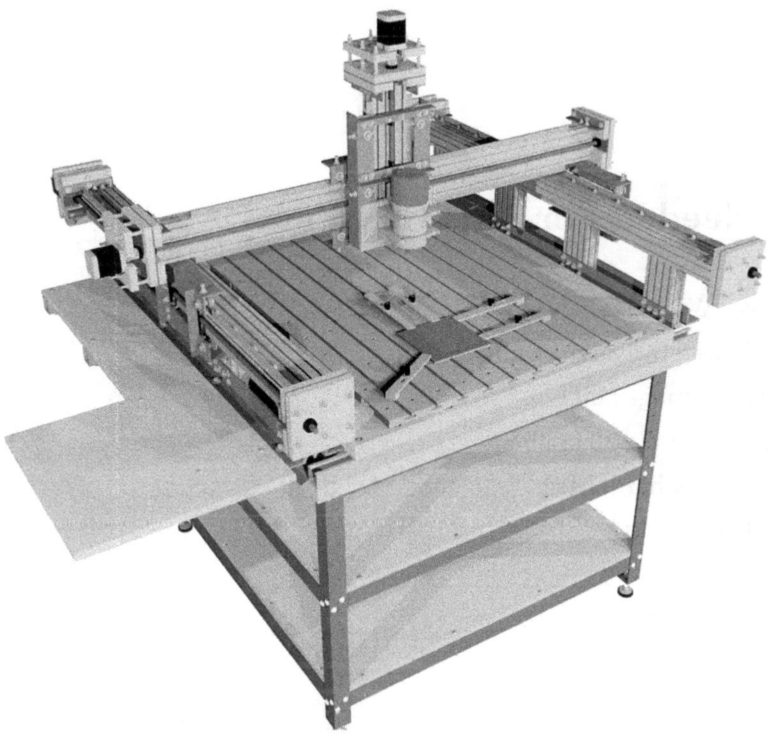

The clamp table serves two purposes. First, it's your sole means of securing the work piece to the CNC. Second, it serves as a auxiliary waster board if your bit happens to protrude through your primary waster board. For this reason, the clamp table must not be hostile to the bit if it comes in contact with it.

Tools Needed For This Chapter

- Hex shank 3/8" countersink and #8 pilot bit
- 3/32" Drill bit
- 1/4" Drill bit
- 1/4" Wood chisel
- Portable drill
- Jig saw, band saw or circular saw
- File, grinder, or rotary grinder

Components Needed For This Chapter

- 1, 24" x 96" x 3/4" White melamine slot wall (see text)
- 100, #6 1-5/8" Coarse thread drywall screws
- 8 1/4-20 x 1-1/2" Carriage bolt
- Incra Build-It System 1/4-20 knob kit
- Some scrap 1/8" thick hardboard.
- Baltic birch (18mm thick)

Chapter Estimates

Cost: $51

Time: One day

Prerequisites

Before staring this chapter you will need a sheet of 24" x 96" white melamine SlotWall, available from Home Depot Part #856307000119. This material is used to create store display systems by allowing one to use special hangers in the provided T-slots. Other home centers will have different names for this material.

This will become the basis for your clamp table. The 24"x 96" sheet will need to be cut into two 43" long sheets as shown in Figures 2.1 and 2.2. The folks at your local home center should be able to cut these for you.

It is important to note that other home centers may carry different sizes of these sheets, making it possible to purchase a full 48" x 96" sheet. If you can get your hands on a full sheet you will only need a single 43" long piece, cut 31-3/4" wide.

Figure 2.1

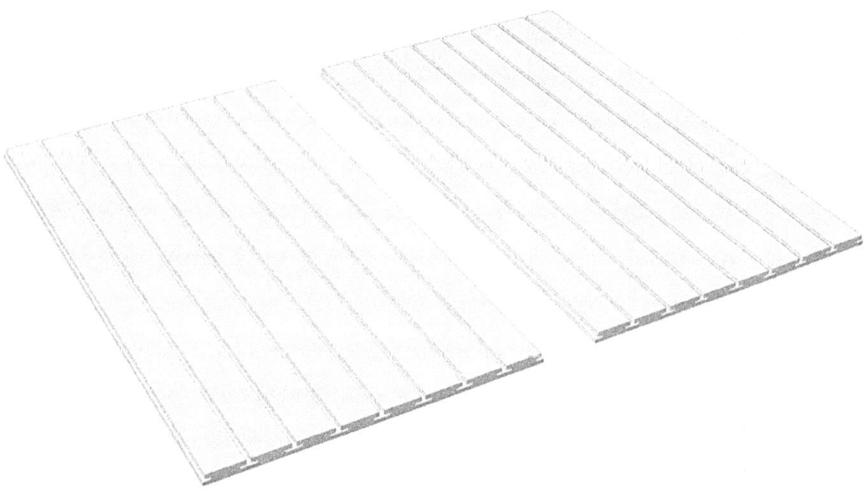

Figure 2.2

Additionally, you are going to need a set of knobs for your clamps. These need to be as small as possible so that they don't get in the way as you rout your parts. The recommended knobs can be purchased from Woodcraft, the kit is called, "Incra Build-It System 1/4-20 Knob Kit". The Woodcraft, part number is #148123. You get eight knobs, bolts, and washers.

Step 1 - Rip Clamp Table Sections to Size

Take one of the 43" pieces of the SlotWall and rip one side 2-1/8" from the edge of the first T-slot, as shown in Figure 2.3. Rip the other side 2" from the edge of the last T-slot. Once cut, the piece should be pretty close to 22-1/2" wide.

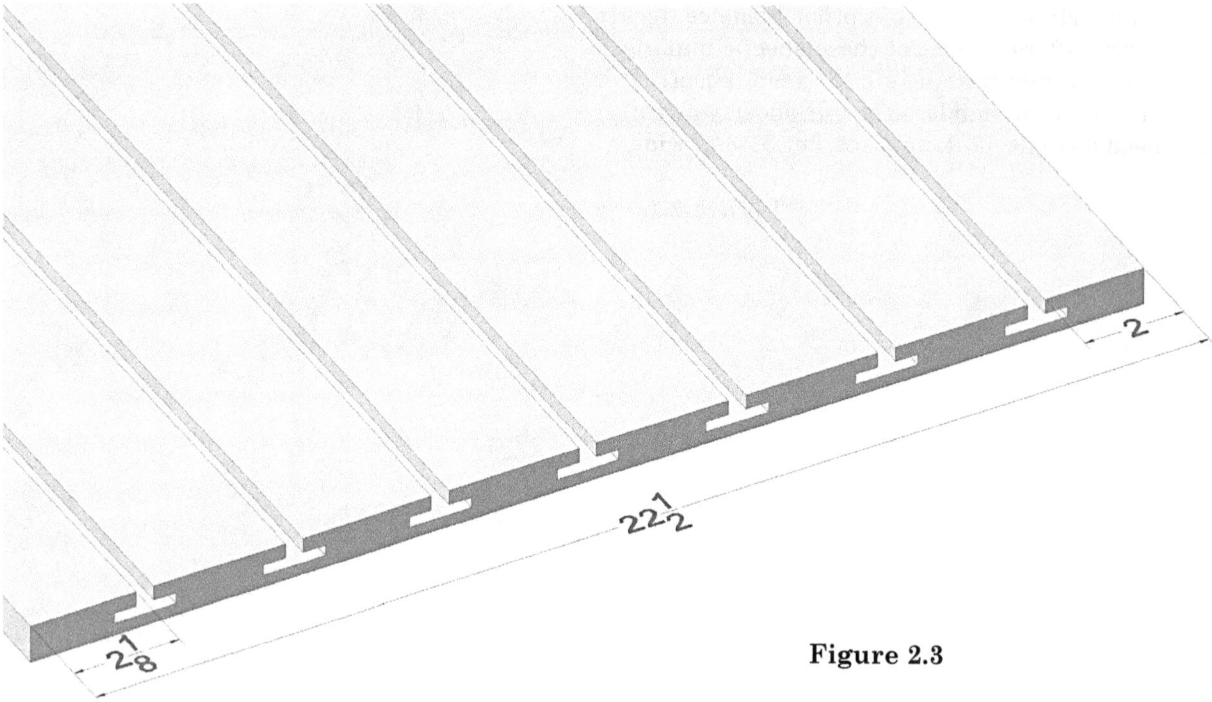

Figure 2.3

Tip

Obviously the best way to rip the material is with a table saw. That said, if you mark your cut lines you can use a jig saw or circular saw to rip the piece. Since nothing is referenced against any of the edges, it wont hurt anything if the cuts are not perfect.

Take the other 43" piece of the SlotWall and rip one side 1-1/4" from the edge of the first T-slot, as shown in Figure 2.4. Rip the other side 1-5/8" from the edge of the third T-slot. Once cut, the piece should be pretty close to 9-1/4" wide.

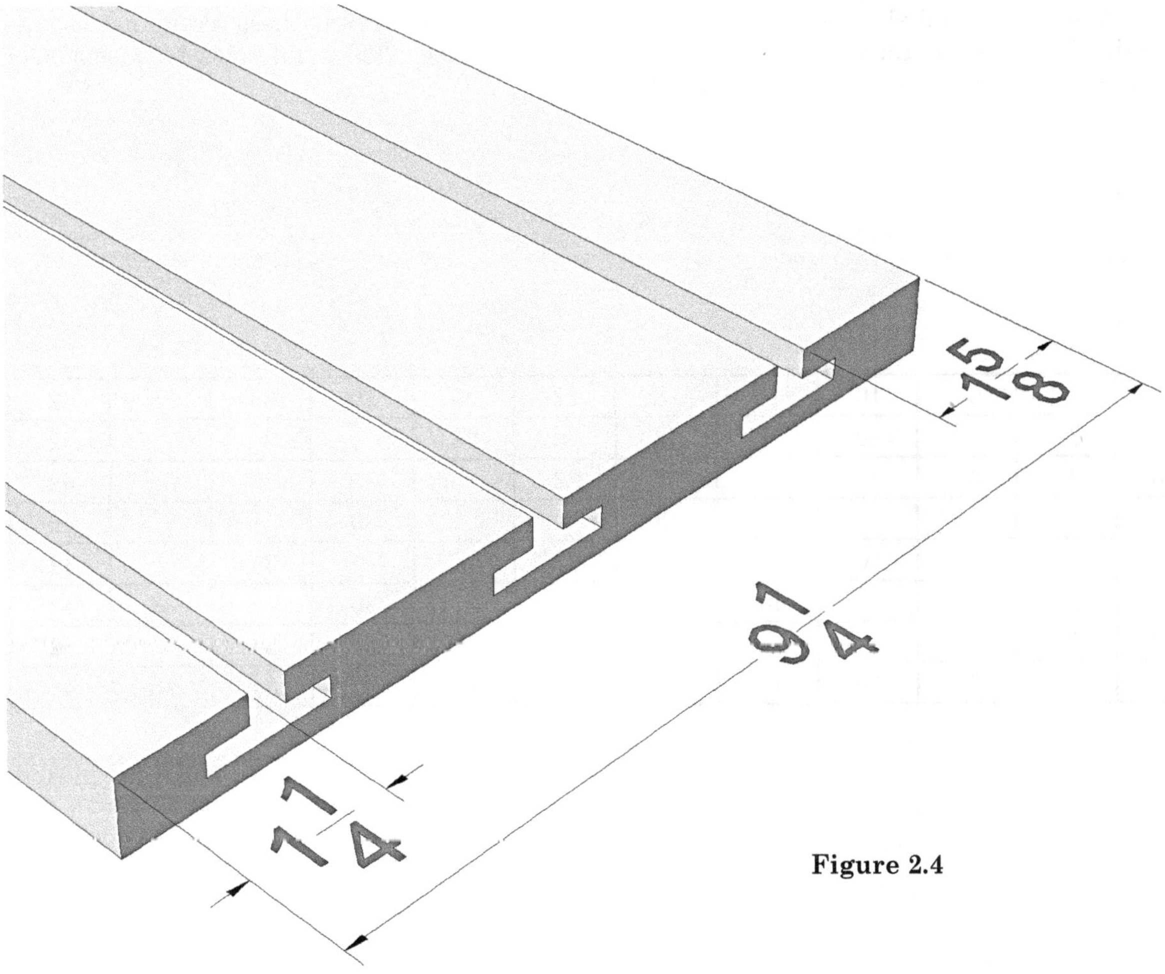

Figure 2.4

Step 2 - Pre Drill Counter Sink Holes

You need to countersink several holes in the larger section of the clamp table. Use the table shown in Figure 2.6 as a guide to the number of holes needed to mount the clamp table. Due to the T-slots, the material has a tendency for the outside edges to roll up, so you will need more screws on the outside as shown. Note that Row 1 and Row 8 are 1" from the outside edge. All other holes are in the center of each section.

The actual placement of the screws is not crucial and the table is only provided as a guide. The holes are 3/8" deep and should also contain an pilot hole for the drywall screws. These screws are used to attach the clamp table to the CNC table top. To drill the holes, use a 3/8" countersink bit with a #6 or #8 pilot bit. These can be purchased at most home centers.

R1	2	6	10	14	18	22	26	30	34	38	42
R2	2		10		18		26		34		42
R3	2	6		14		22		30		38	42
R4	2				18			30			42
R5	2		10			22			34		42
R6	2	6		14		22		30		38	42
R7	2		10		18		26		34		42
R8	2	6	10	14	18	22	26	30	34	38	42

Figure 2.6

Take a piece of masking tape and mark off the 3/8" point on the countersink, as shown in Figure 2.5. When you drill the hole, stop when the tape reaches the surface of your stock. Figure 2.7 shows the large section of the clamp table with the holes drilled.

Figure 2.5

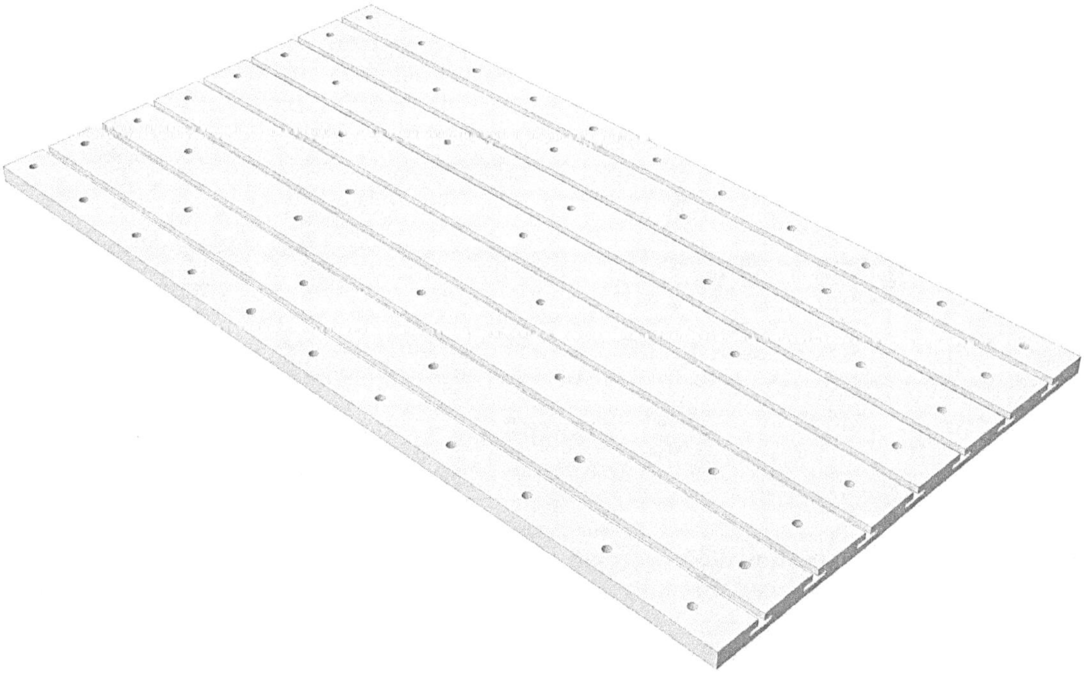

Figure 2.7

Take the smaller clamp table section and mark and drill the holes shown using the table shown in Figure 2.8. Drill them in the same manner as you did with the large section. Figure 2.9 shows the finished section of the clamp table.

R1	2	6	10	14	18	22	26	30	34	38	42
R2	2	6		14		22		30		38	42
R3	2		10		18		26		34		42
R4	2	6	10	14	18	22	26	30	34	38	42

Figure 2.8

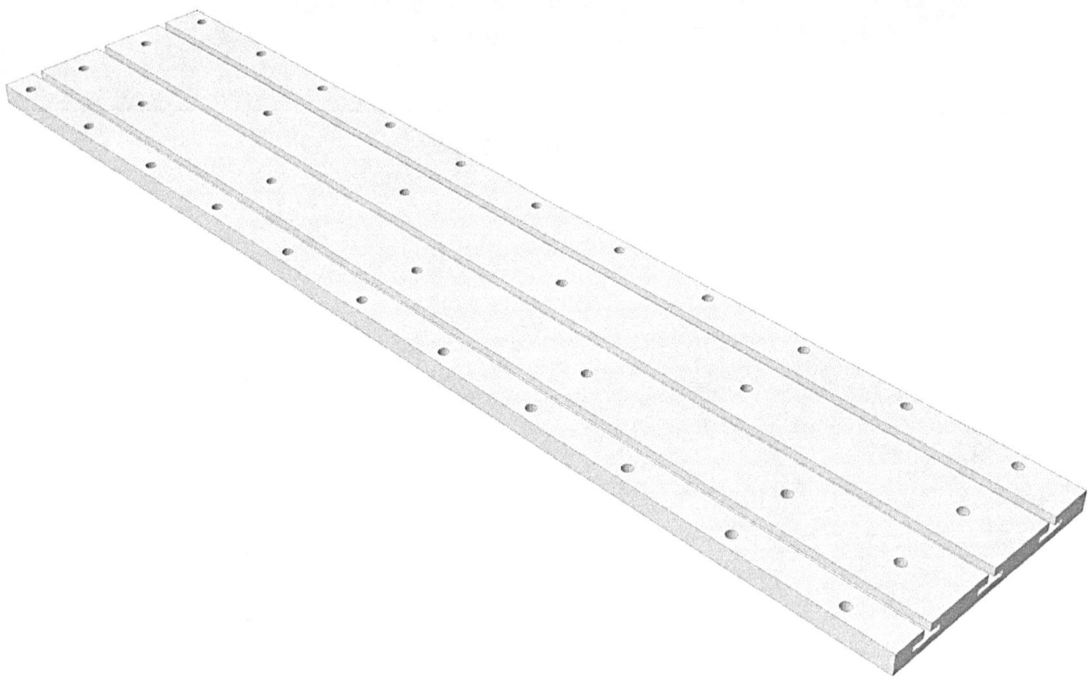

Figure 2.9

Once all the holes have been drilled, you will need to clean the bottom side of each clamp table section where the pilot bit punched through. It's important that the holes are clean and that all debris has been removed.

Step 3 - Attach Clamp Table Sections to CNC

Lay the larger clamp table section on the KRMx01 table as shown in Figure 2.10. The end with the holes 2" from the edge should be facing the front. The front left corner should be 4-3/16" from the right edge of the KRMx01 table and 3/4" from the front edge of the KRMx01 table. Once positioned, mark all four corners as you will be removing the section.

Once marked, use a 3/32" drill bit to run the pilot holes into the KRMx01 table at the four corners. Then, hold the section corners firmly in place against the table and attach it with four #6 drywall screws. With the corners attached, finish drilling the rest of the pilot holes. Don't attach any more screws at this point.

When you drill the pilot holes, debris will get in between the clamp table sections and the KRMx01 table. Back out the four screws in the corners and remove the clamp table section. Use a vacuum to thoroughly clean the KRMx01 table and the bottom side of the clamp table section.

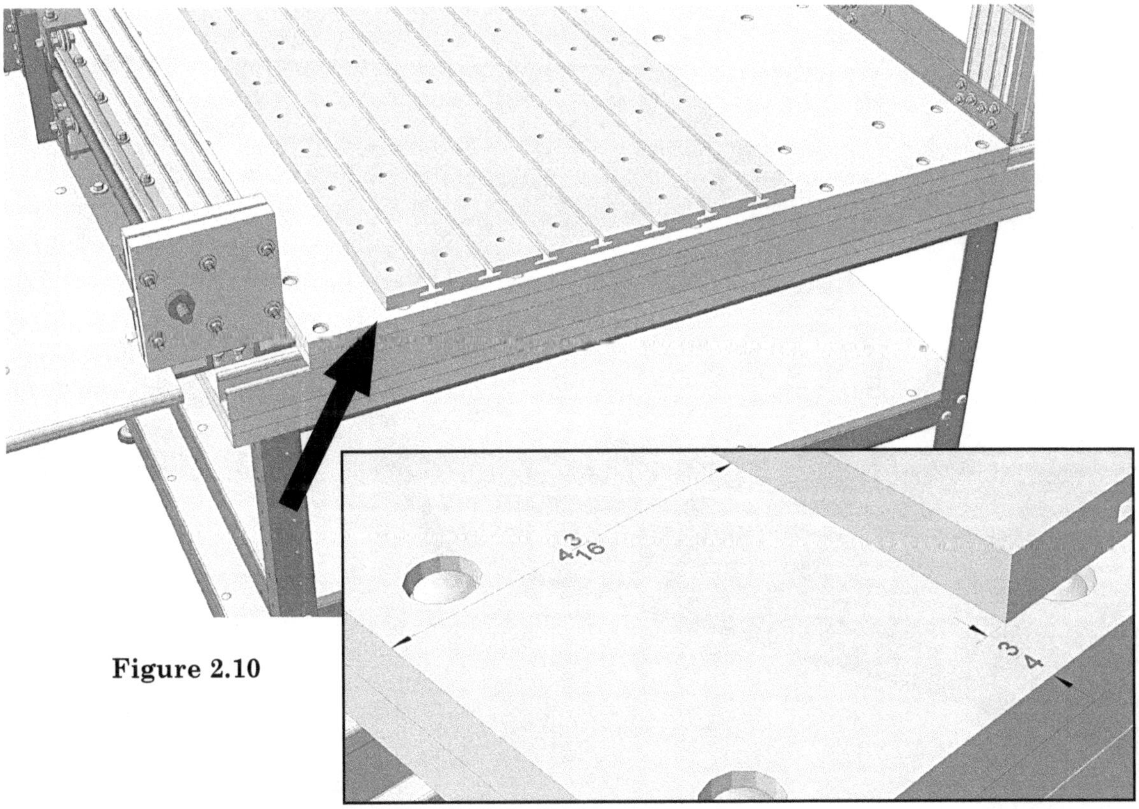

Figure 2.10

Place the clamp table section back on the CNC. While pressing firmly on the clamp table, attach the first row to the CNC by using 11, #6 drywall screws. It is extremely important that the clamp table be firmly set against the KRMx01 table when tightening the screws.

Next, attach the last row with 11, drywall screws. Proceed with the eight screws on the front and back edge of the clamp table and work your way into the center attaching the screws.

Tip

You can actually get up on the top of your CNC and use your knee to hold the edge in place while you attach the screws.

Set the smaller clamp table section in place by butting it against the larger one, as shown in Figure 2.11. Repeat the process of drilling the pilot holes, cleaning them, then reattaching the section.

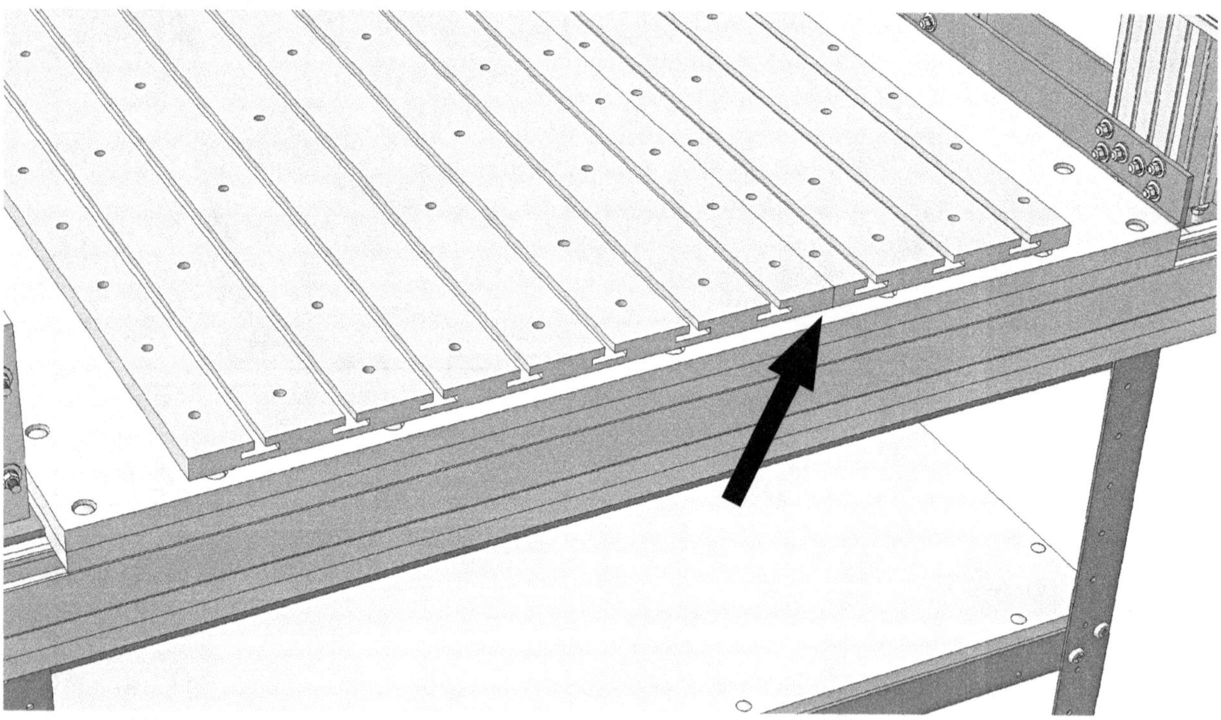

Figure 2.11

Step 4 - Make the Clamp Bushing

In order to properly use the clamp table, you need to make clamps. The heart of the clamp is the clamp bushing shown in Figure 2.12. It can be made from hardboard, plastic, or even metal if you prefer.

Use the layout shown in Figure 2.13 to cut a piece of 1/8" material 1-1/2" x 1-1/8". Place a 1/4" square hole in the middle. The hole does not have to be perfect, so you can use a variety of methods to make the hole. The simplest is by using a hammer and chisel.

Once complete, round over the corners with a file or sander.

You will need about 10 of these, made manually. Later, you will use the CNC to make more of the clamp bushings.

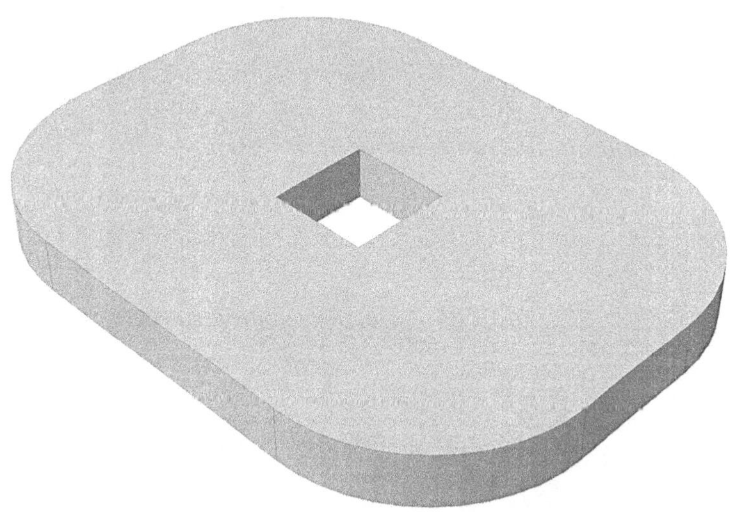

Figure 2.12

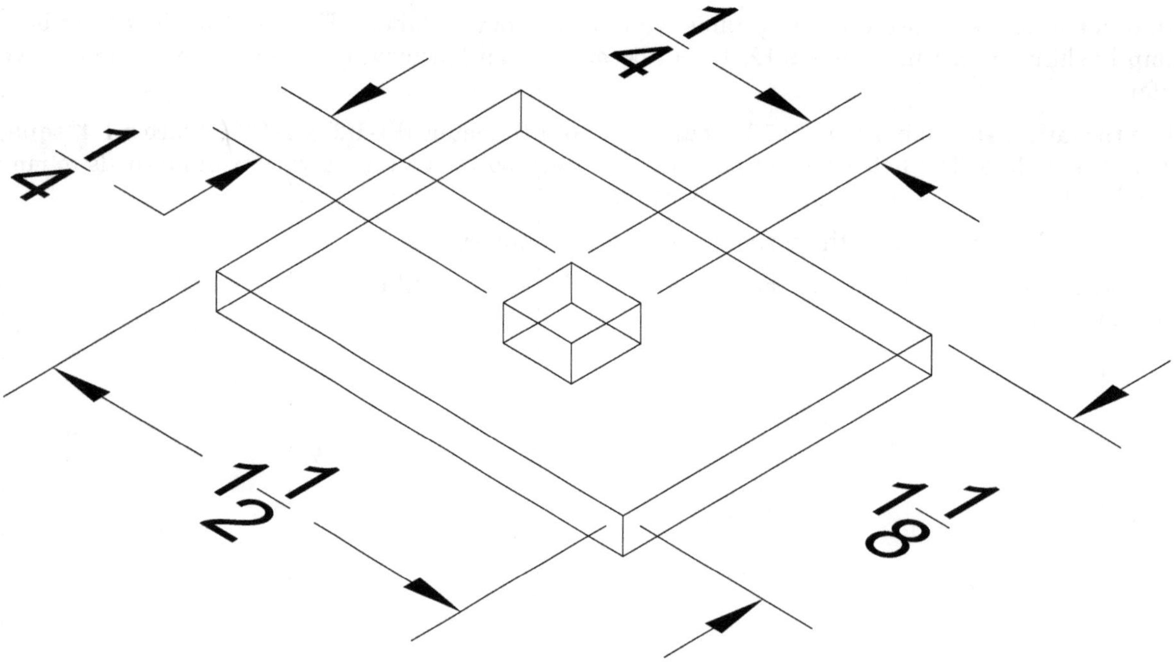

Figure 2.13

Tip

The easiest way to make the bushing is to create a 1-1/8" wide strip out of hardboard, then cut them into sections 1-1/2" long.

To cut the 1/4" square hole, drill a 1/4" hole in the middle with a drill bit and then use a 1/4" chisel to square it up.

Step 5 - Prepare the Clamp Bolt

You will be using a 1/4-20 x 1-1/2" carriage bolt that will be pushed through the center hole in the bushing. Unfortunately, the head of the bolt is slightly thicker than required and needs to be filed down to slide smoothly through the T-slots.

To achieve this, use a bench grinder, or rotary tool to remove approximately 1/32" of material from the top of the bolt head, as shown in Figure 2.14.

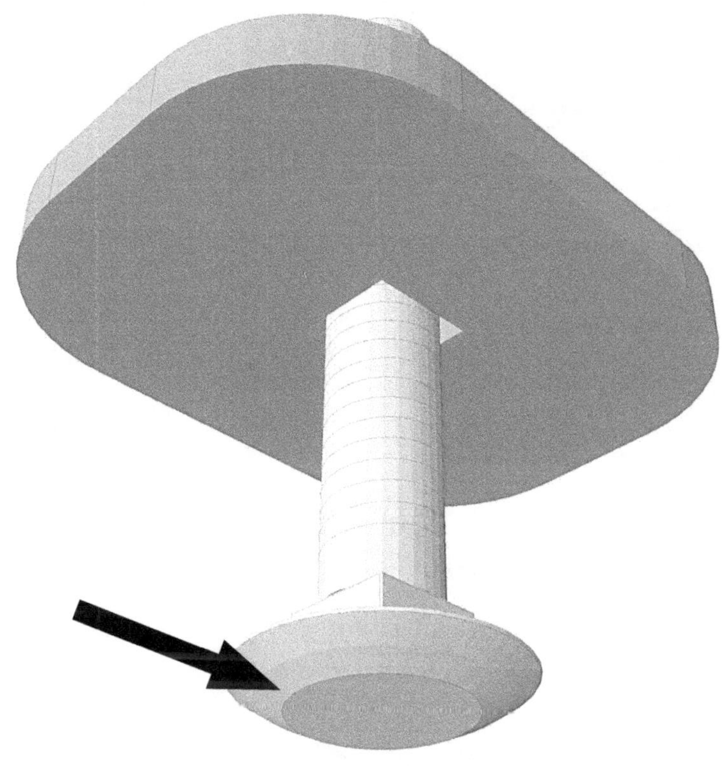

Figure 2.14

Step 6 - Make the Clamp

Let us take a detailed look at a clamp that's been designed to work with the KRMx01 CNC. The clamp is made from a piece of Baltic birch 8-1/2" x 1" x 11/16" thick. It has a 7-3/4" x 7/16" section cut out of the body and a 1/8" x 1/8" section taken off the tip, as shown in Figures 2.15 and 2.16. There are two elongated holes in the top for the clamp bolt to protrude. Probably the most important aspect of the design is the 1/8" section taken off the tip. This creates a small rabbit that will allow you to overhang your stock, thus securing it from the top and the side at the same time.

It is important that the clamp be made from Baltic birch; the many plies add strength and rigidity while still allowing the router bit to survive if it contacts the clamp.

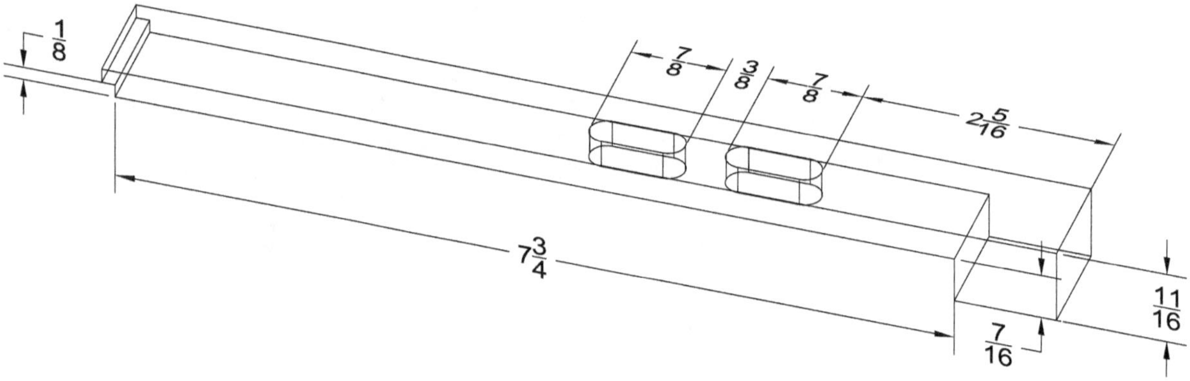

Figure 2.15

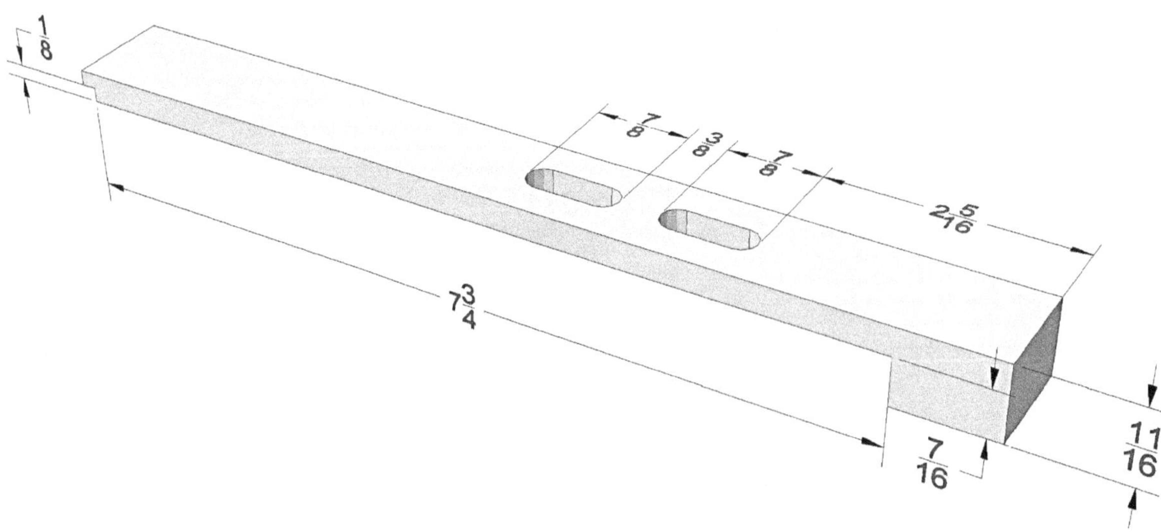

Figure 2.16

Cut the 8-1/2" x 1" strips on a table saw, then use a band saw or scroll saw to cut the profiles. The slots in the top are 7/16" wide. Their placement is not critical.

You can use a drill press to drill a series of 7/16" holes along the length of the clamp. You will find that as you use the clamps, you may need to add additional holes.

Take a look at the clamp in Figure 2.17. It's a single 1/4" strip of wood about 9" long and has a small slot chiseled into the end, with a couple of holes added to the top. You can pick up molding strips at your home center that will work for this type clamp.

Use any kind of saw to make the slots or . You can also use hand tools such as a coping saw or chisel. The finished clamp should look like the one shown in Figure 2.17. You can also use a drill press to drill single holes, then use a chisel to combine them.

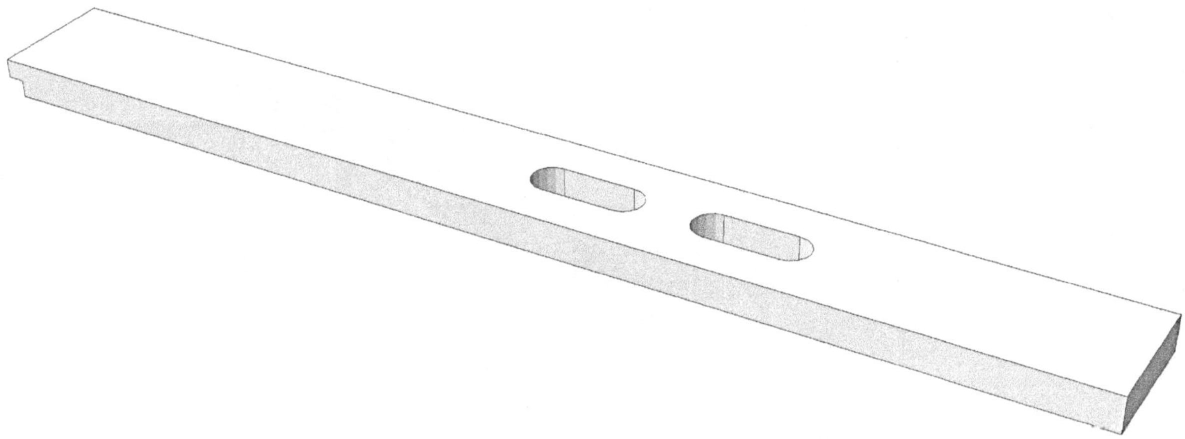

Figure 2.17

Step 7 - Attaching the Clamp to Clamp Table

The process of attaching the clamp assembly to the clamp table is a simple one.

Start by inserting the prepared carriage bolt into the clamp busing. Then insert it into one of the T-slots as shown in Figure 2.18. If it does not slide in smoothly, you will need to grind a little more off the head of the bolt until the bolt assembly moves freely.

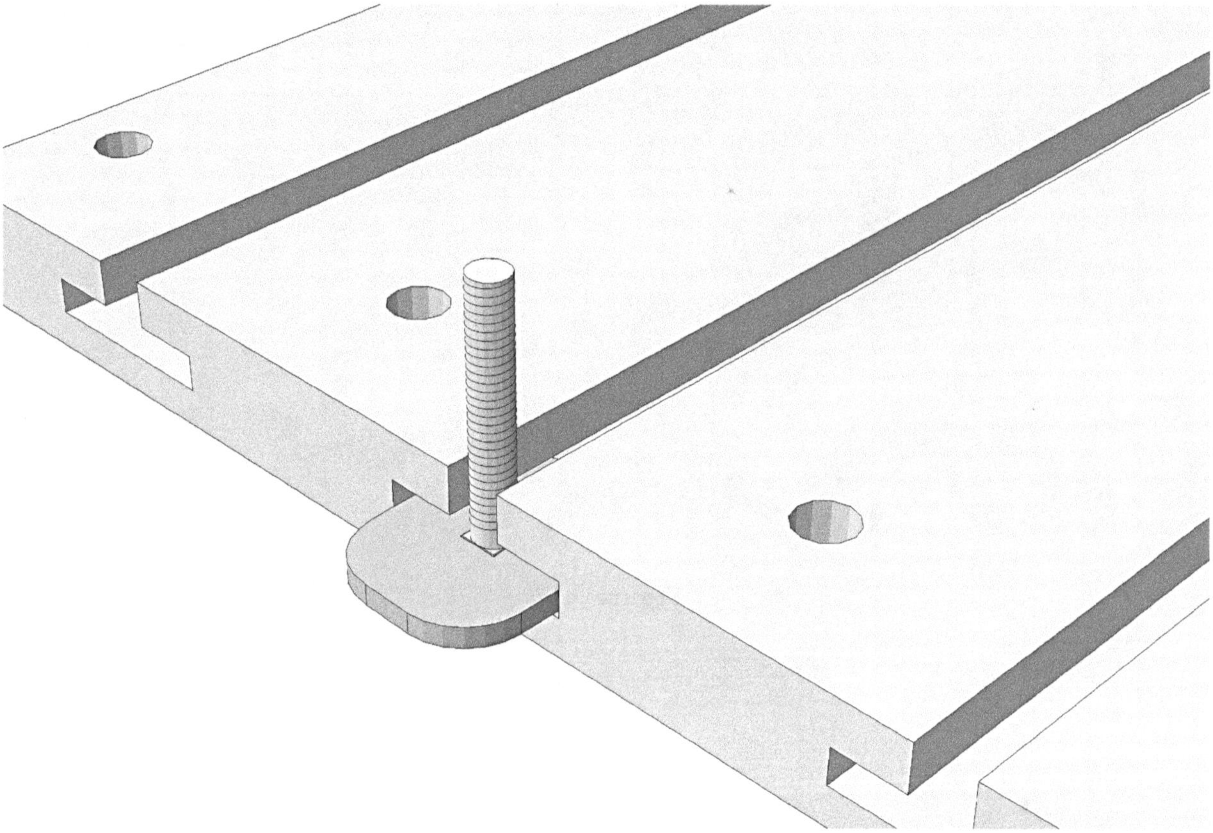

Figure 2.18

Next, drop the clamp over the bolt as shown in Figure 2.19. The actual hole you use will be dependent upon the slot and stock size. Remember, you want to try and keep the bolts and knobs as far away from the stock as possible.

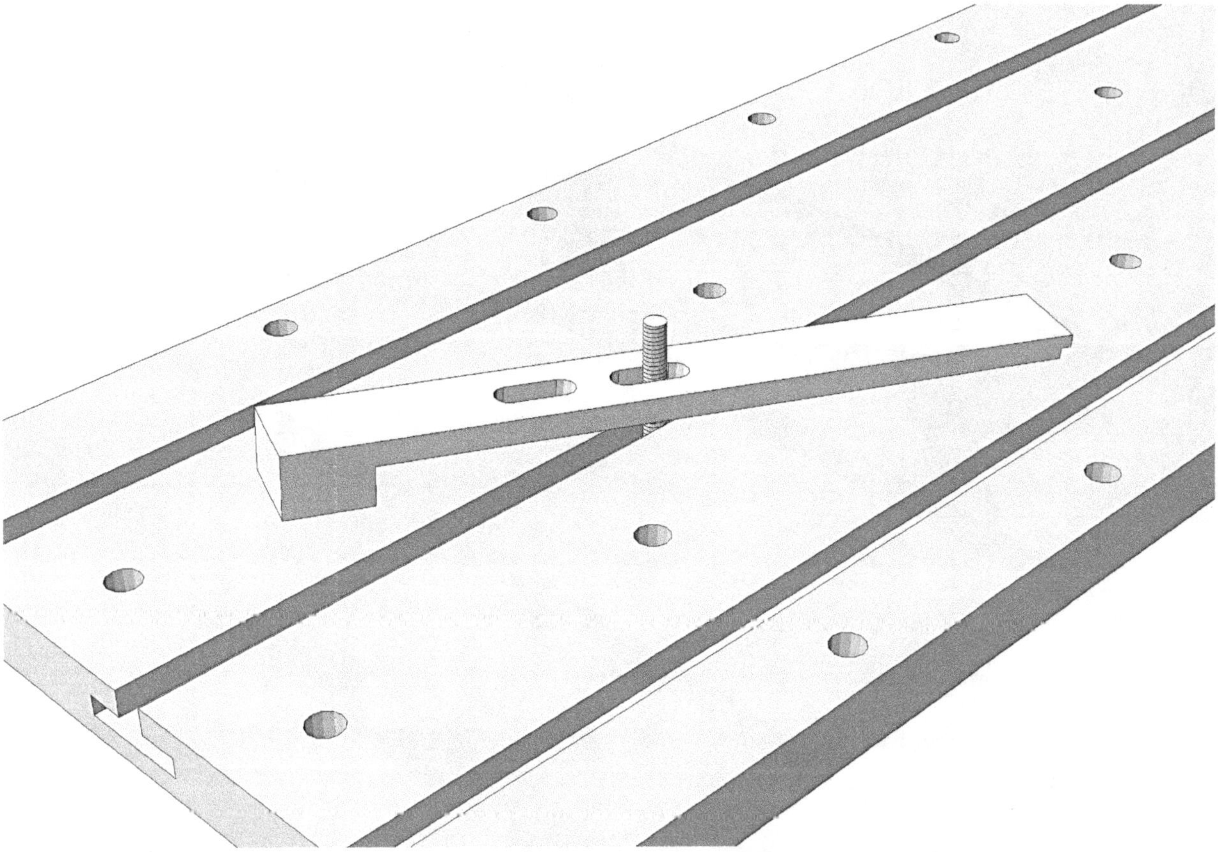

Figure 2.19

Finally, insert the washer and knob and tighten as needed to secure the stock, as shown in Figure 2.20.

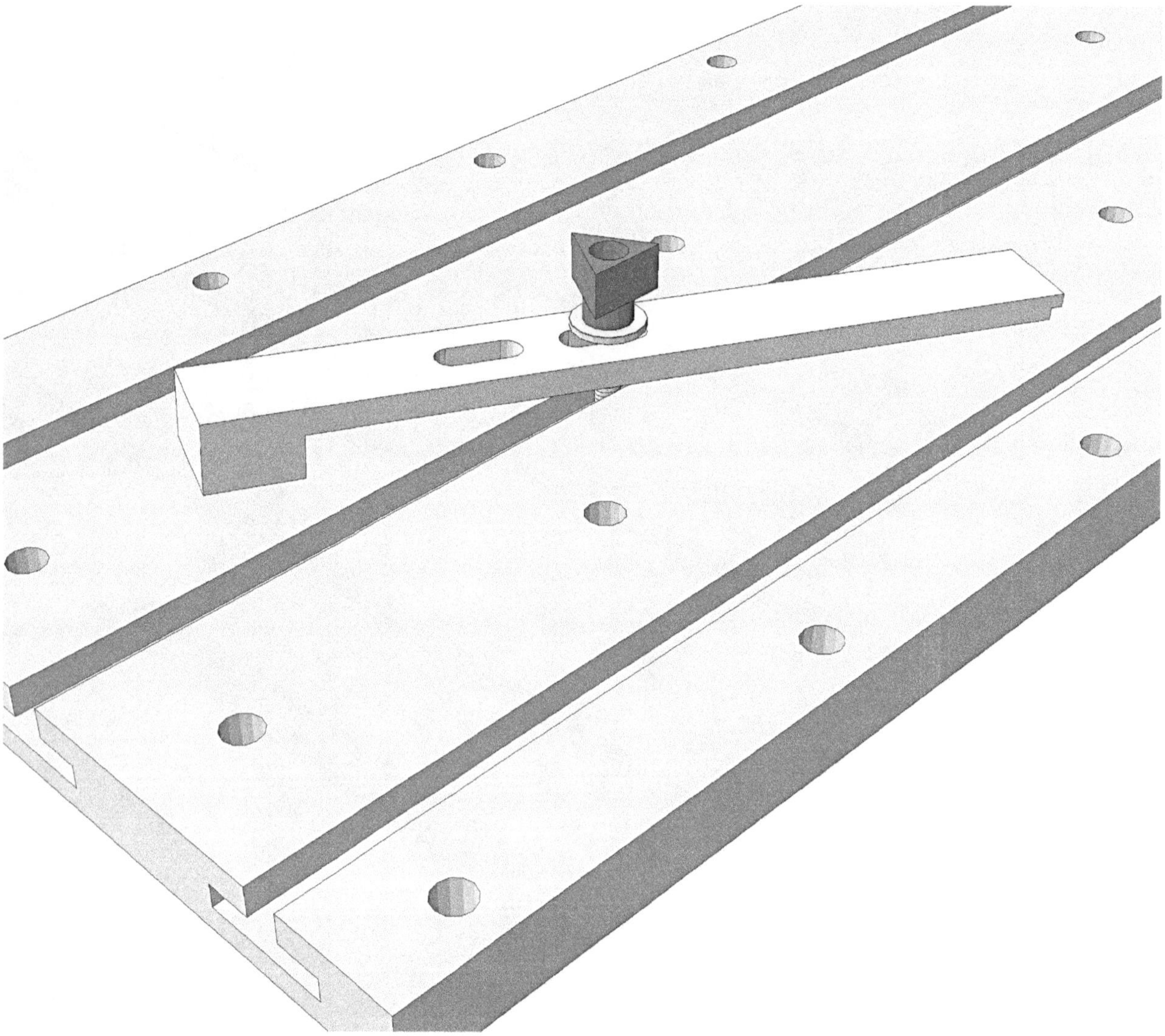

Figure 2.20

Clamp Table In Use

Most of the time you will be holding two pieces of stock in place.

The lower piece is waster board. Its sole purpose is to allow you to cut through the top piece of stock without damaging your clamp table.

Note that there are eight clamps shown in Figure 2.21. With this configuration, the stock is held in place by the top pressure of the clamp. More importantly, the lip provides side pressure that keeps the bit from moving the stock from side to side.

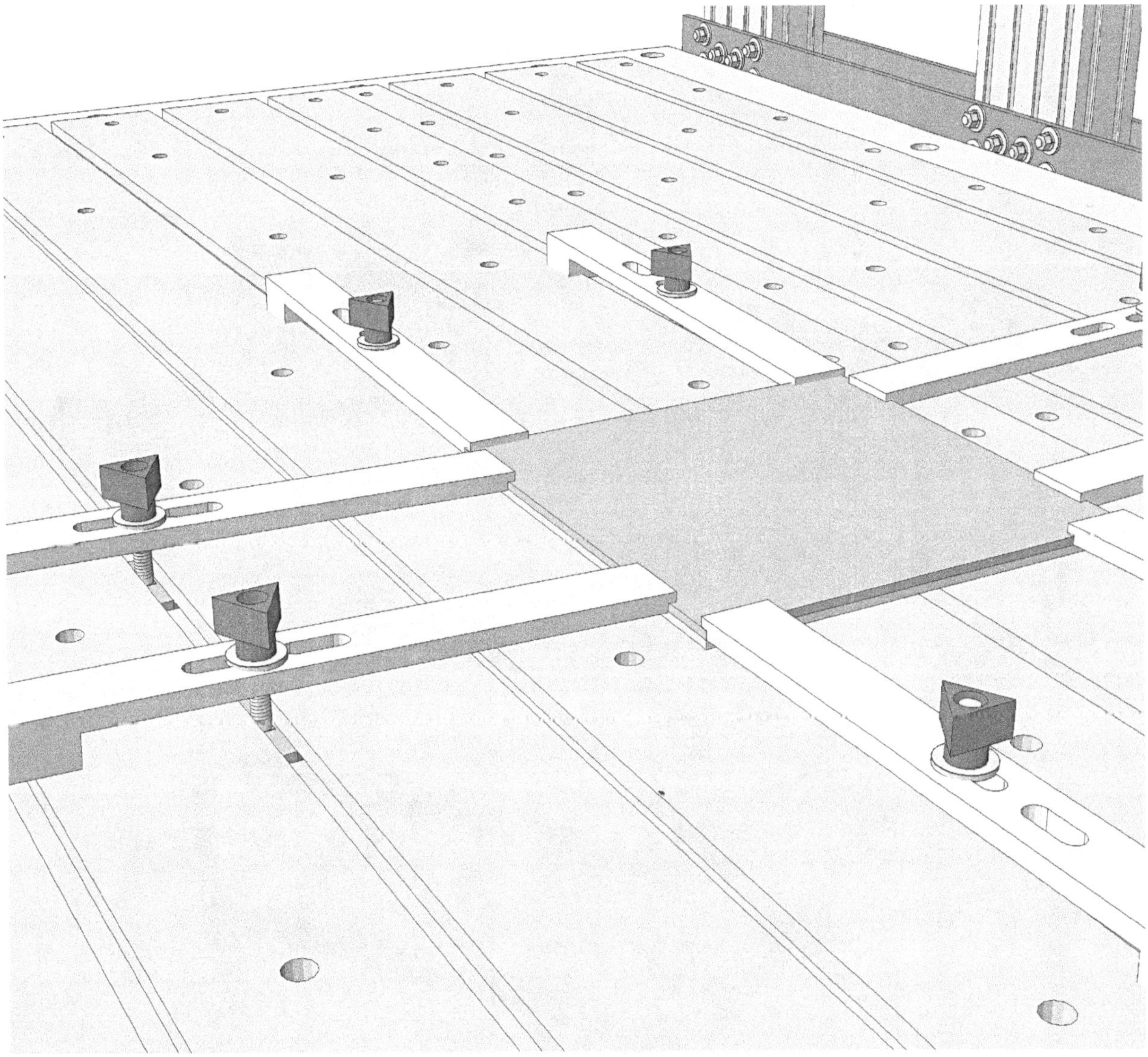

Figure 2.21

Conclusion

Now that your clamp table is in place, you need only add a router to begin machining material. In the next chapter, you will add fences to your CNC. These will allow you to setup your stock much faster. It will also insure consistent milling from part to part.

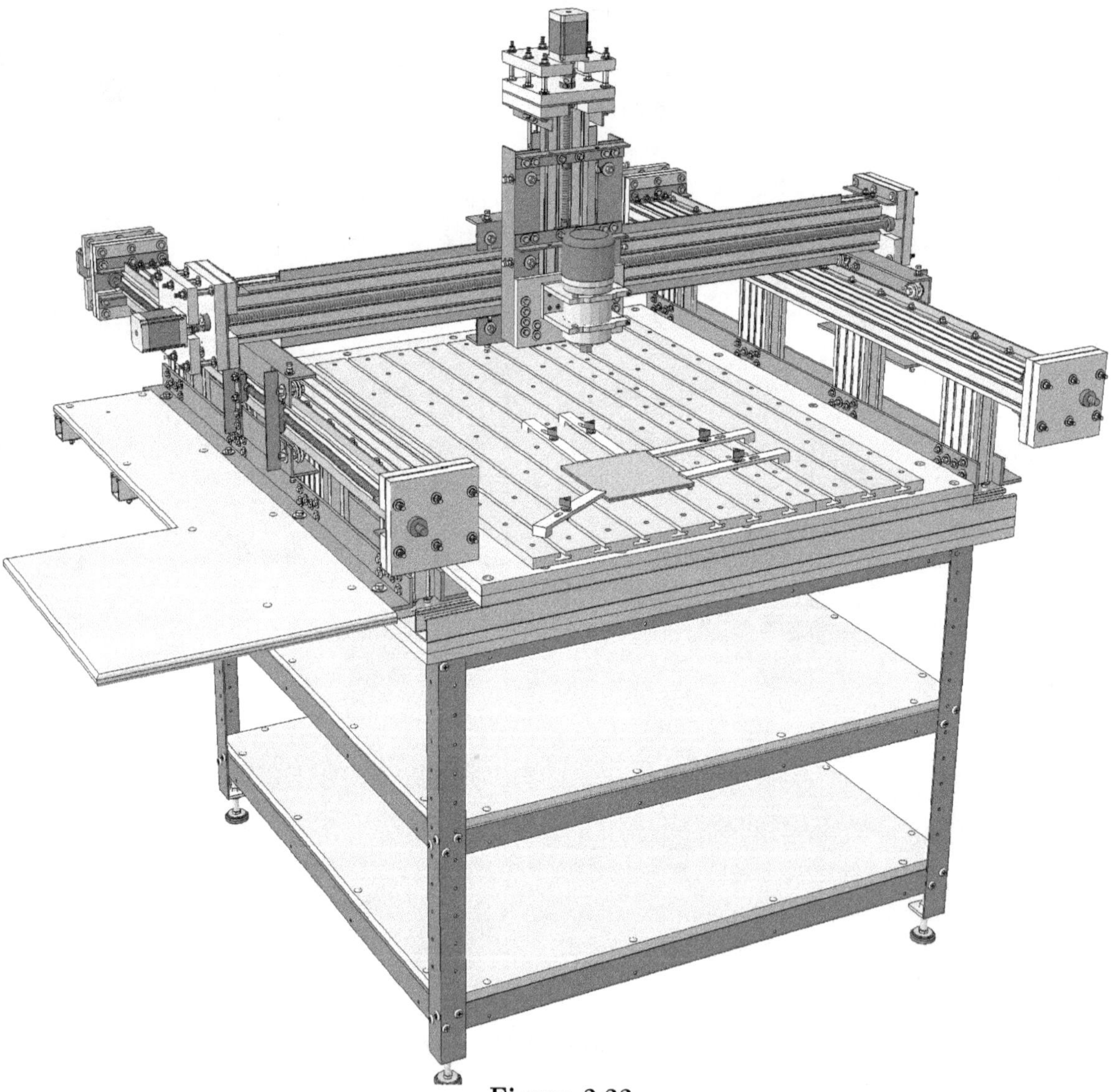

Figure 2.22

Chapter 3

Fences

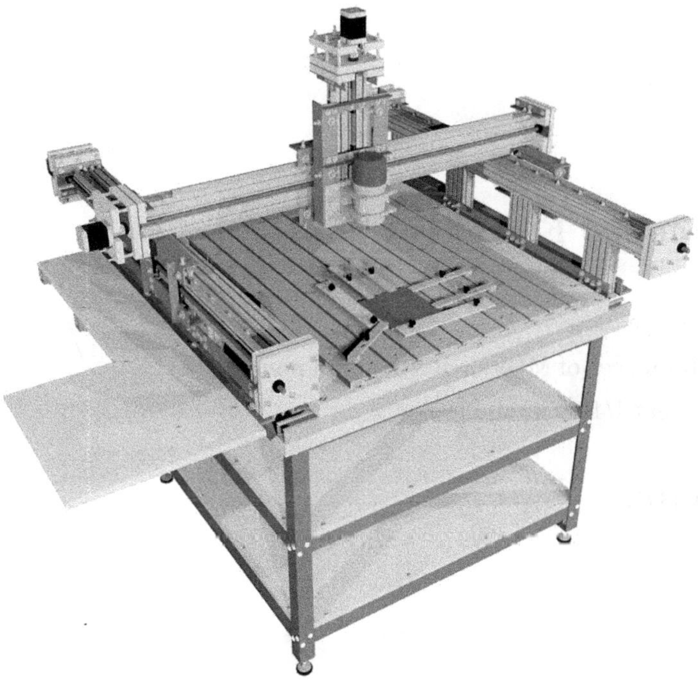

If you are a woodworker, you are already familiar with fences. Most of your stationary tools have them. The use of a fence on your CNC can make it very easy to go from part to part without a lot of setup.

In this chapter, you will learn how to add two small 12" fences that can be used anywhere on the table.

Tools Needed For This Chapter

- Saw used to cut the fence material to size (table saw, chop saw, band saw)
- Framing square

Components Needed For This Chapter

- 4, 1/4-20 x 1" Carriage bolts
- 2, 12-3/4" x 2" x 1/4 pieces of MDF or Baltic birch plywood
- Incra Build-It System 1/4-20 knob kit

Chapter Estimates

Cost: $15

Time: Couple hours

Prerequisites

For this chapter, two 12-3/4" x 2" x 1/4" pieces of MDF or Baltic Birch plywood have been used, although you can use any material that is 1/4" thick as long as it won't damage a router bit. In other words, don't use metal.

Additionally, you can use just about any saw to cut these pieces as long as one side is smooth. If you do use a hand held power tool to cut the width and length, place the cut edge towards the outside of the CNC, leaving the uncut side towards the work piece.

You will also need four clamp bushings described in the last chapter.

Tip

Feel free to paint your fences to accent your CNC. Make sure the paint is dry before mounting them on your clamp table.

Step 1 - Cut the Mounting Holes in Fences

The two fences should have already been cut to size. Place two 5/16" holes in the center of each fence at the points shown in Figures 3.1 and 3.2. Hole size and actual placement is not critical as long as the two holes are 9" apart.

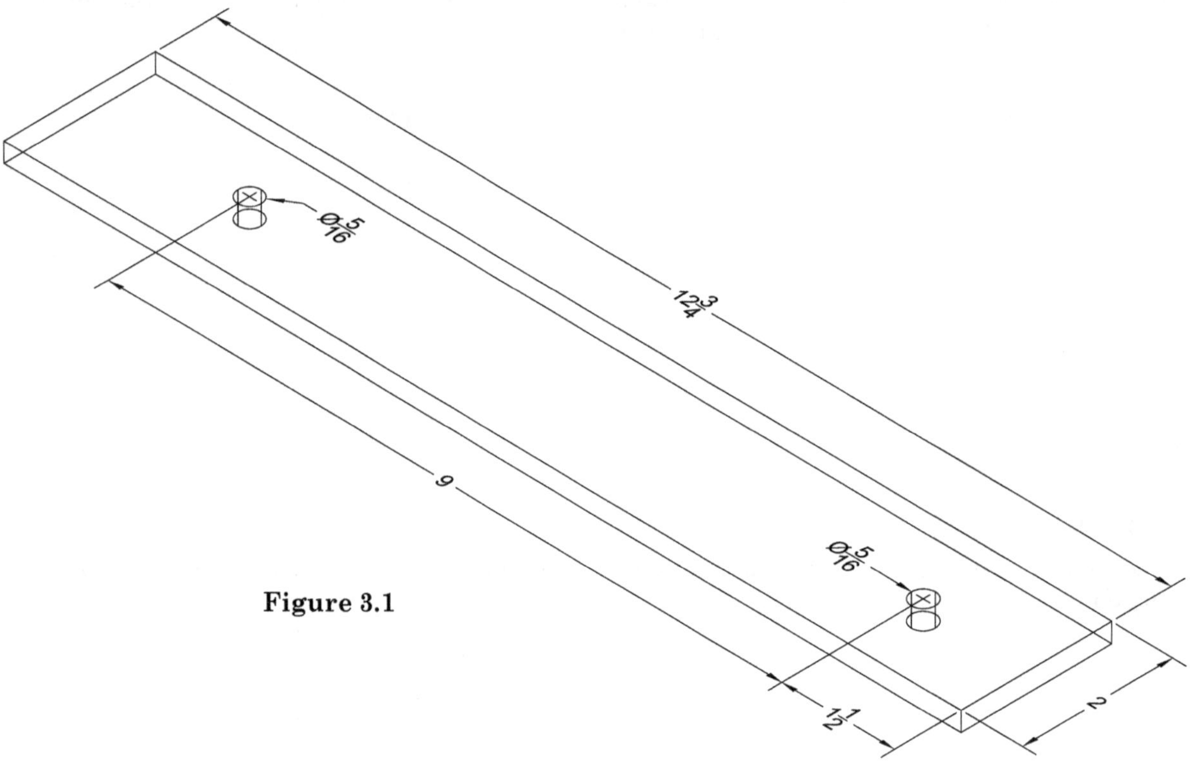

Figure 3.1

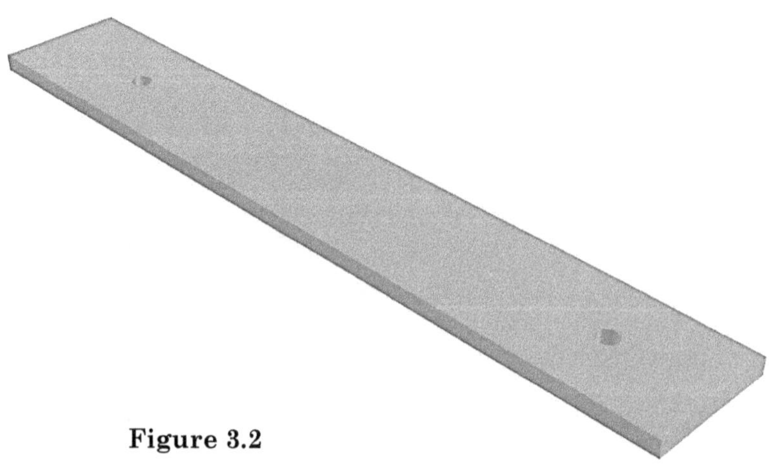

Figure 3.2

Step 2 - Mount Fences to Clamp Table

Insert a 1/4-20 x 1" carriage bolt into one of the clamp busings. Then, insert the bolt into one of the holes on the fence. Add a washer and knob.

Place the two fences in the slots as shown in Figure 3.3. The fence closest to the front should be about 3" from the edge. Notice the small gap between the two fences. This gap is important as it gives the bit a place to go when zeroing the bit on the stock. This topic will be explained in better detail in the next chapter.

Figure 3.3

Step 3 - Aligning the Fences

Insert a 1/4" straight cutting bit into the router. A straight flute bit works best here. Using the arrow keys, move the bit so that one of the cutting edges just touches the front edge of the side fence as shown in Figure 3.4.

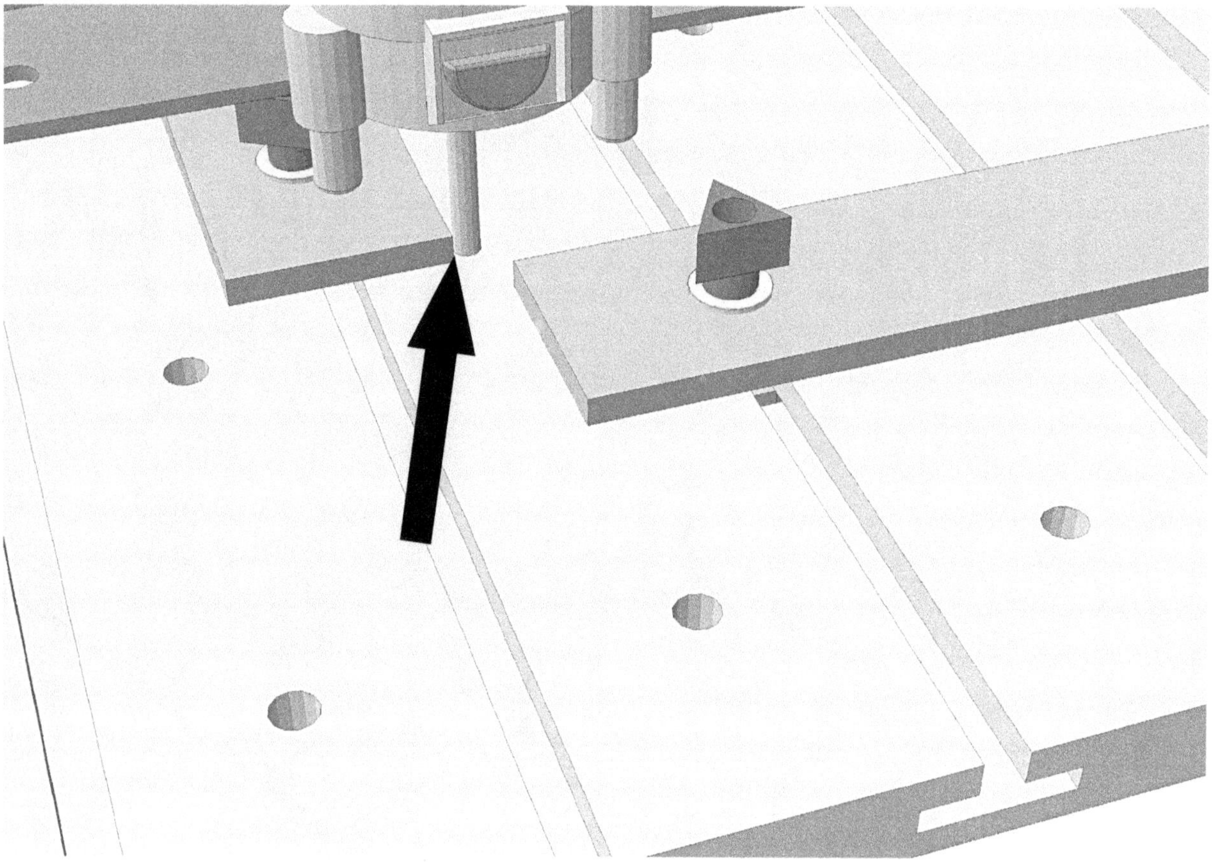

Figure 3.4

Raise the router so that the bit clears the fence and move the router to the rear of the fence as shown in Figure 3.5. Slightly loosen the rear knob and slowly lower the router. Move the rear of the fence as needed so that the bit clears the fence.

Once the bit is lowered, move the fence up against the edge of the bit and tighten the knob.

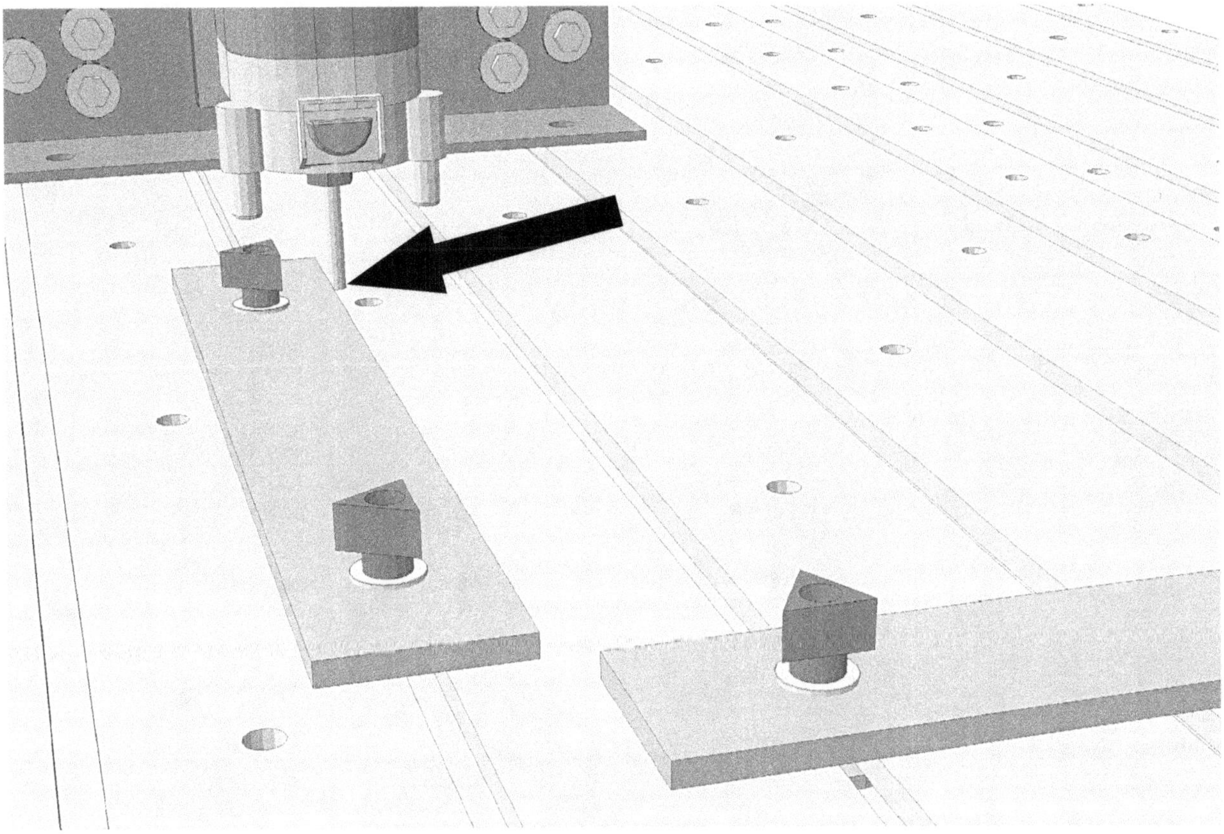

Figure 3.5

Tip

It is important that you keep the bit in a straight line as you move it to the rear of the fence. You want the back of the fence to hit the bit at the same point as it did in the front.

You can purchase blank 1/4" rods for doing this kind of alignment. Many router template kits come with an alignment bit.

The final step in aligning the fence is to adjust the front fence so that it is at a right angle to the side fence. Use a square as shown in Figure 3.6 to perform this operation.

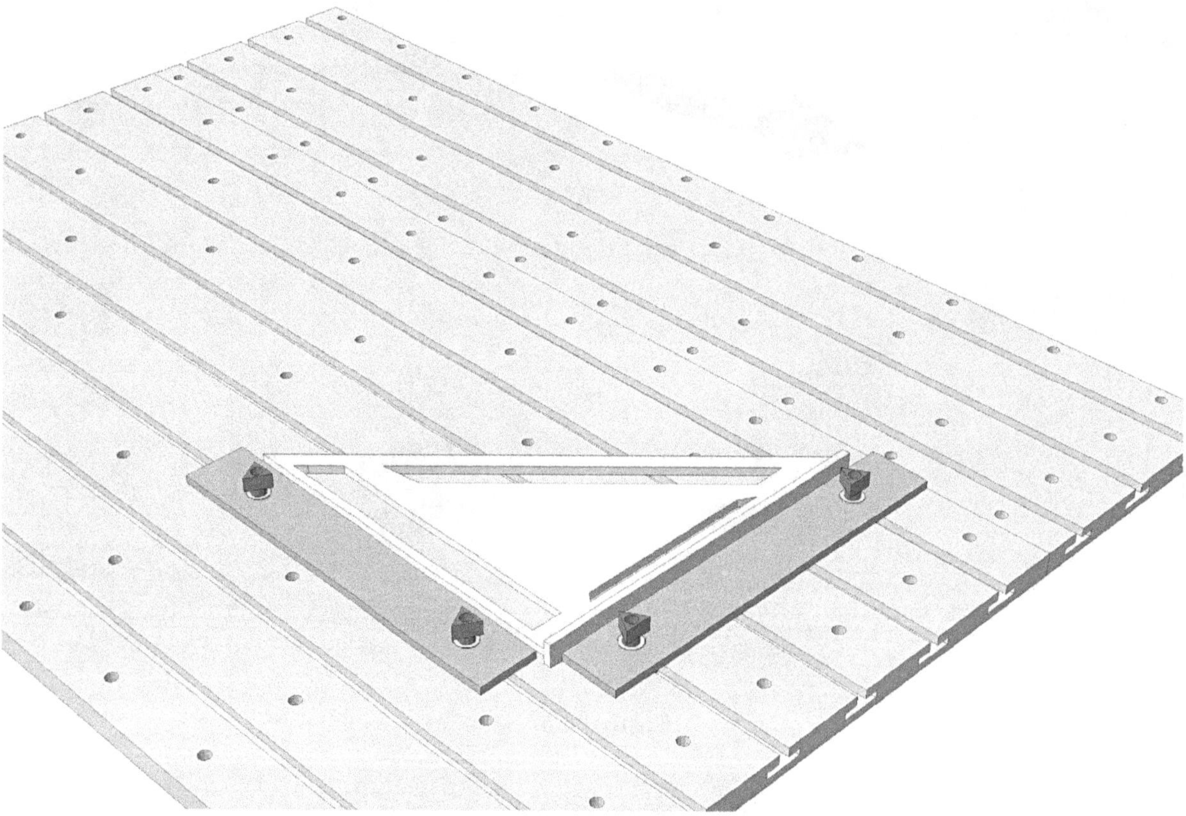

Figure 3.6

Step 3 - Example Clamp Layout

Figure 3.7 shows a piece of 7" x 7" stock positioned against the two fences using five clamps to hold it in place. It is important to note that you will be using many types of clamp and fence layouts. No one layout will work for every part. There are a few questions you should ask yourself when mounting a piece of stock:

- Does the bit clear all the clamps?
- Does the bit clear all the knobs?
- Do the clamps keep the piece from raising?
- Do the clamps keep the piece from sliding?
- Is there a properly sized piece of waster board under the stock?

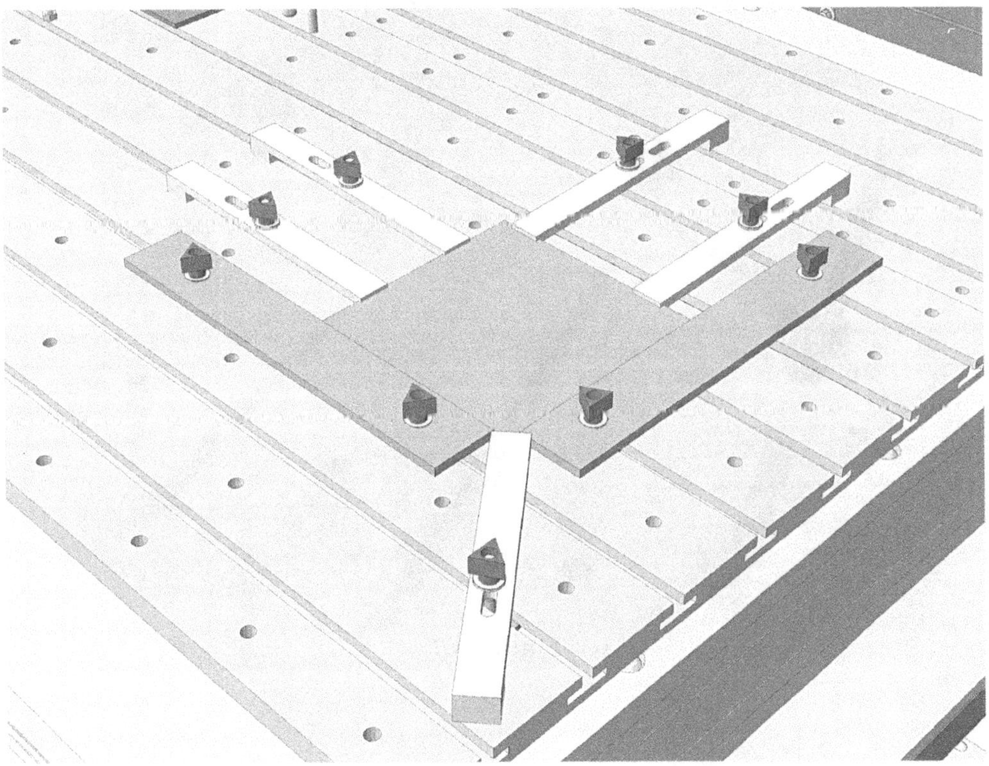

Figure 3.7

Adjustments

Note that the router bit may not move perfectly along the edge of the front fence until the Y axis has been properly adjusted. This adjustment can be done manually or automatically using homing switches.

The instructions for adding homing switches can be found later in this book, for now, alignment must be done manually.

To manually adjust the Y axis, you need to tweak the difference between the Y axis (left beam) and the A axis (right beam). This can be done by removing power to the steppers and manually turning one lead screw or the other.

Figure 3.7

Chapter 4

Basic Operation

In this chapter, you will learn how to use your CNC machine to make an improved clamp bushing similar the ones you created when you installed your clamp table.

This new clamp bushing has a small pocket that holds the carriage bolt. You no longer need to "Prepare" the carriage bolts; any 1/4-20 carriage bolt can be used as-is.

Tools Needed For This Chapter

- 1/8 Spiral router bit (see text)
- Saw used to cut stock and waster board (table saw, chop saw, band saw)

Components Needed For This Chapter

- 3 pieces of 7" x 7" x 1/8" hardboard

Chapter Estimates

Cost: $10

Time: Couple hours

Important Notice

This chapter involves operating your CNC without an emergency stop switch, which is an upgrade featured later in this book. Without this feature, you will have to hit the reset button on Mach 3, and if something goes wrong you could damage your router bit. A properly installed emergency stop switch does not guarantee this won't happen, but it can give you a quicker response time.

In the past you could have used Mach 3 in demo mode to do basic testing and simple cuts on your CNC. This project has more than 500 instructions so you will need to purchase a Mach3 licence if you have not already done so.

Prerequisites

At this point, you should have your KRMx01 CNC up and running. Your cables should be arranged so that you can machine a part in a 12" x 12" area without them getting caught or tangled. You should have the clamp table and at least five clamps completed. Your two fences should be installed and adjusted.

You need to cut three 7" x 7" x 1/8" pieces of hardboard, as shown in Figure 4.1. One of these will be used for your waster board. The other two will be used to make parts. If you wish, the waster board can be made from scrap material.

Figure 4.1

You will need a 1/8" spiral router bit like the one shown in Figure 4.2. You can use an up spiral or a down spiral. The up spiral will give you a better finish on the bottom of the piece, and the down spiral will result in a better finish on the top of the piece. With this project, either will work just fine.

If you need to order these bits online, use the following Freud part numbers:

- Freud Up Spiral Bit # 75-100
- Freud Down Spiral Bit # 76-100

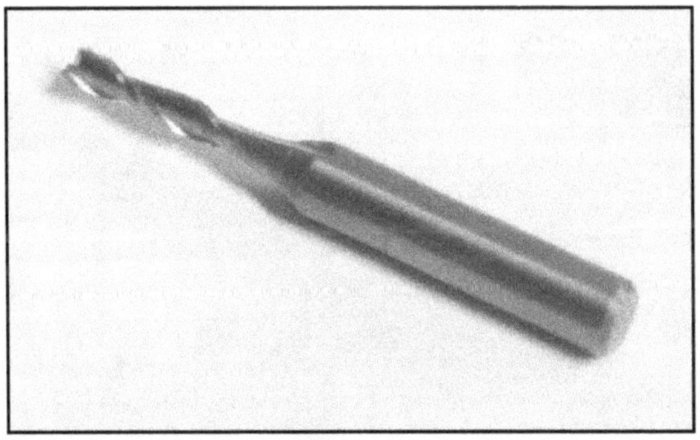

Figure 4.2

Tip

As you cut various shapes and sizes of waster board, save them. You will find that you can use them many times before replacement. Want to know where to store them? How about the upper shelf on the CNC stand?

Another name for waster board is spoil board.

Step 1 - Clamp Stock to the CNC

Start by stacking one of the 7" x 7" x 1/8" pieces of hardboard onto your 7" x 7" x 1/8" waster board and mounting both against your fences as shown in Figure 4.3. Use two clamps on each of the exposed sides. The clamp in the front left corner should be slightly offset, exposing the corner .

Figure 4.3

Step 2 - Prepare the KRMx01

Power up your PC and start the Mach 3 software by clicking on the KRMx01 shortcut.

Load your spiral bit into your Router. You need to have about 1-1/4" of the bit protruding beyond the router collet.

Power up your G540 controller by plugging in your 48V power supply.

The green LED on the G540 should be lit.

Step 3 - Position the Router

Set the jog speed to 10% and use the arrow keys to move the tip (center) of the router bit to the corner of the work piece, as shown in Figures 4.4 and 4.5. Use the page up and page down keys to set the bit exactly on the surface of the piece. You want the corner of your stock to be dead center of the router bit.

Note: Remember, you need to make Mach 3 active by hitting reset

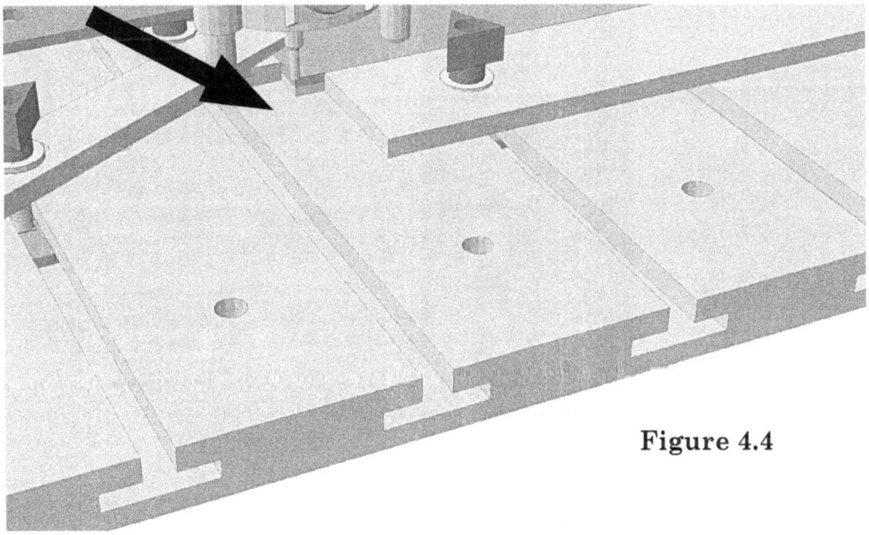

Figure 4.4

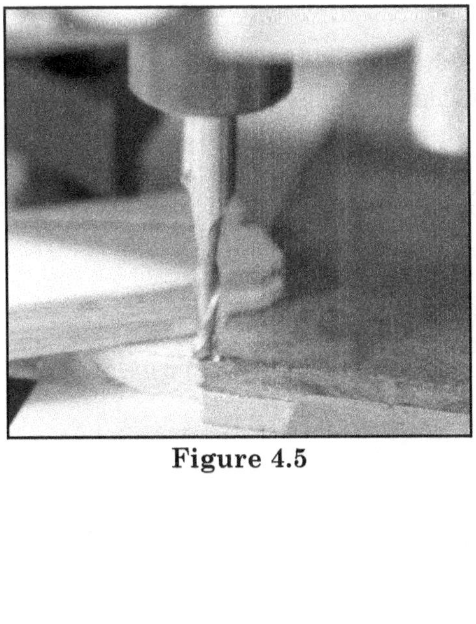

Figure 4.5

Tip

To set the jog speed, hit the tab button to bring up the jog control. Use the + and - pads to set the slow jog rate.

Step 4 - Set the KRMx01 Reference Point

The CNC needs to have a point of reference so that it can accurately machine a part at a know position. To do this, move to a known location and tell Mach 3 where that position is.

When you start Mach 3, the four DRO's should look something like the ones shown in Figure 4.6. These numbers most likely will be what they were the last time you closed the Mach 3 application.

The red outline on the Zero buttons are telling you that you have not yet referenced your machine.

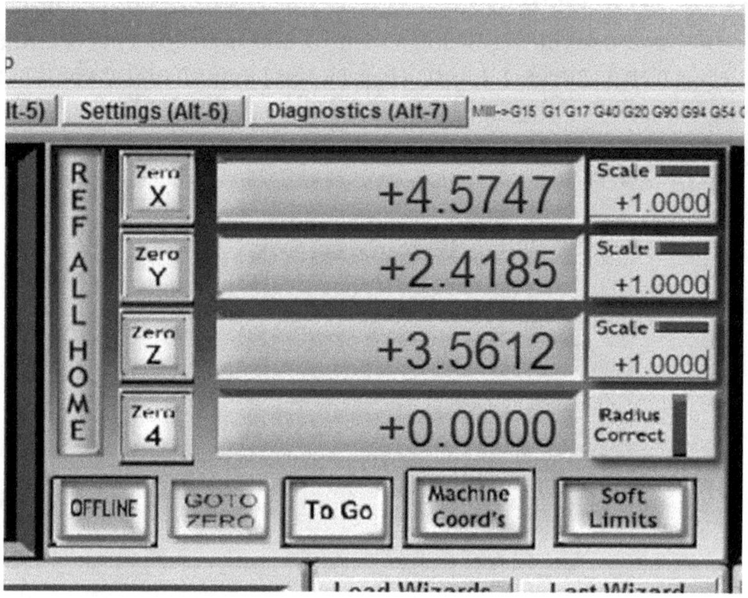

Figure 4.6

Tip

DRO stands for Digital Read Out. These are the large green fields on the display.

With your router bit moved to the position indicated in step 3, and using Figure 4.7 as a guide, hit the large button labeled "REF ALL HOME." Hit each of the Zero buttons. The DRO's should display zero and the outline on the buttons should glow green.

At this point, your CNC will call this place home. All movements will be based on this position being at x=0, y=0, z=0.

Any time your CNC loses step you will need to reposition your home point and re-reference Mach 3.

What causes a CNC to lose step?

- Powering off the G540
- Hitting a limit on any axis
- Crashing the bit into an object
- Moving faster than the router bit can cut

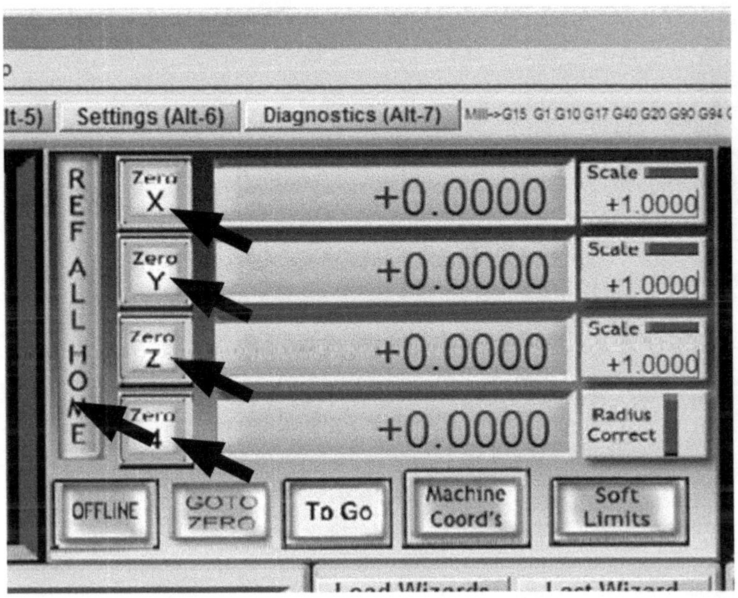

Figure 4.7

Step 5 - Load G-code File

G-code is a series of commands that tell Mach 3 how to control your CNC. They tell the CNC where to go and how fast to go. They can turn on the router (future upgrade). They can even set the speed of the router (another future upgrade).

To load a G-code file, select the Load G-code option from the File Menu as shown in Figure 4.8. This will load the Open File dialog box.

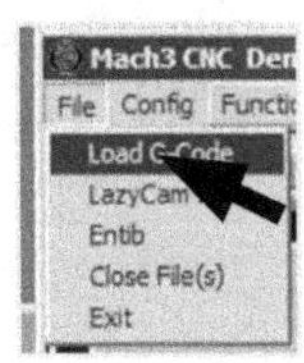

Figure 4.8

Download the clamp bushing G-code from the Kronos Robotics web site. The file is called "Single Clamp Bushing.txt." Once downloaded, select that file and hit the Open button, as shown in Figure 4.9.

You can download the file here:

http://www.kronosrobotics.com/krmx01/da69781_9386_87112/

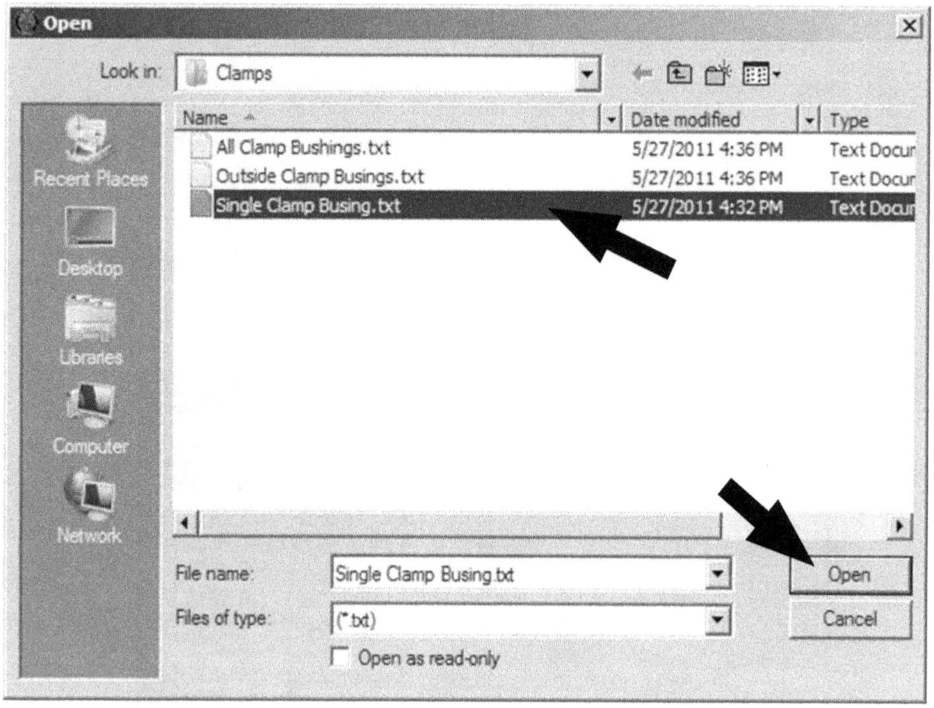

Figure 4.9

Tip

You are going to perform a dry run by moving the router bit up and re-referencing the home position. You can also run through the program by placing the software offline. This removes the charge pump from the G540, effectively shutting down the motors. If you do take the CNC offline, you will have to re-reference your machine again when it is brought back online. The offline button can be found under the reference button shown in Figure 4.10.

Figure 4.10

When the G-code file has been loaded you should see the instructions in a large green box on the left hand side of the display as shown in Figure 4.11. In the large black box on the right hand side of the display, you should see a graphical representation of the actual tool path.

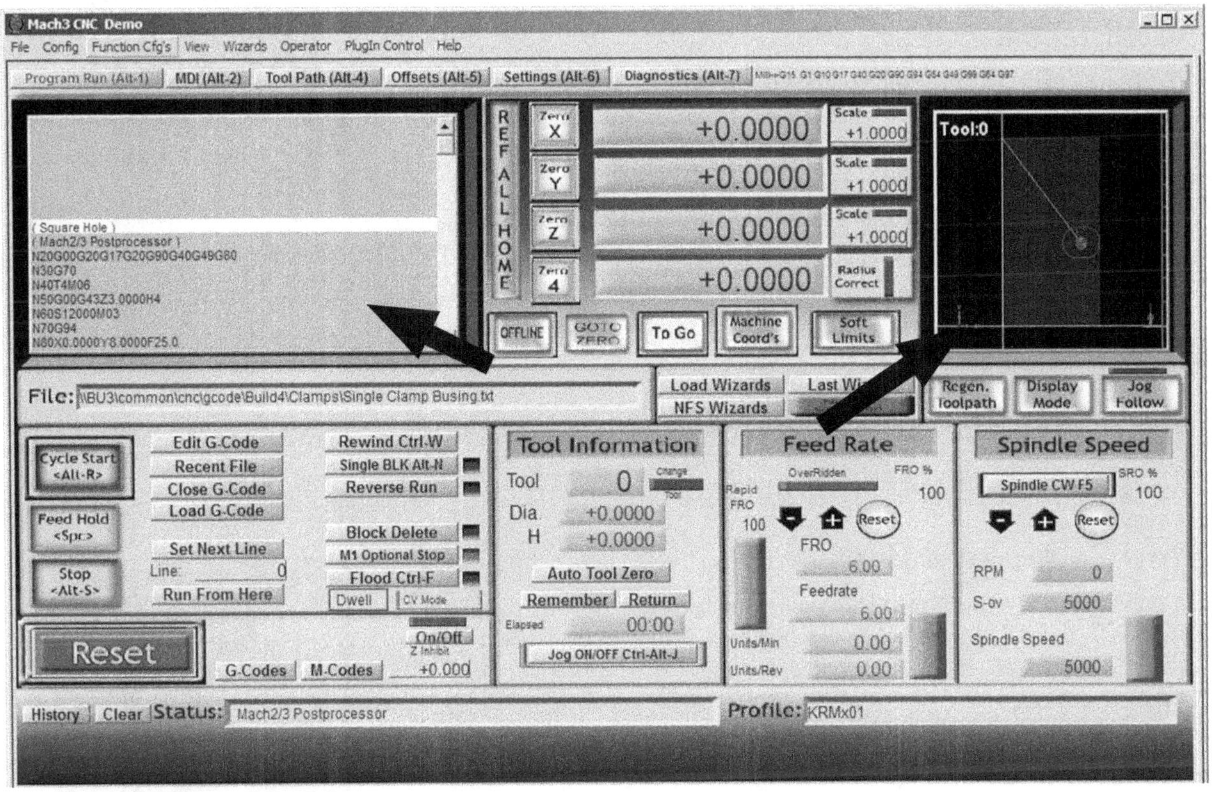

Figure 4.11

Step 6 - Perform a Test Run

Until you get used to the operation of your CNC and develop some sort of workflow, I recommend performing test runs. To perform a test run, do the following:

1. Move your already referenced machine to the home position. (0,0,0).
2. Use the page up key and move the bit approximately 1" above the surface.
3. Hit the Zero Z button.

Your machine is now referencing a position 1" higher than your stock. To run the G-code, hit the Cycle Start button shown in Figure 4.12. We are not going to turn the router on at this point so hit the Cycle Start button now.

Figure 4.12

You should see the router go through its paces about an inch above your material. It should be near the center of the material.

Since we have not wired an emergency cut off, you need to be prepared to hit the reset button if it looks like your router is going where it's not supposed to go.

Step 7 - Make a Part

You did a test run and all went as expected. The G-code program file should have automatically rewound and should be back at the top of the instructions as before.

Hit the GOTO ZERO button shown in Figure 4.13, then manually move your Z axis back down to the surface of your stock and hit the Zero Z button.

Figure 4.13

Once referenced and all zeros are showing in your DRO's, move the Z axis approximately 2" above your surface. Get into the habit of never starting a program or router with your Z axis resting on your material.

Set the speed of your router to 4, and turn on the router.

Start the program by hitting the Cycle Start button shown in Figure 4.14 and let the program run through its instructions. The "Single Clamp Busing" has over 500 steps and should take about 53 seconds to run.

Figure 4.14

Once the router has finished, it should move itself up and out of the way. Turn the router off and use a shop vacuum to clean your piece. It should look like the one shown in Figure 4.15 and the edges will be a bit frayed, which is normal. Don't remove the piece as you are going to cut the rest of the clamp busings.

If the part has been cut into the center of the stock then you know you are on track. There also should be a small pocket cut into the center of the clamp bushing that is approximately 1/16" deep.

Figure 4.15

Step 8 - Make More Parts

Close the current G-code file by selecting the Close File option from the File menu as shown in Figure 4.16.

Download the file called "OutSide Clamp Bushings.txt". As before load the file into Mach3 using the Load G-code menu option. Your machine should still be referenced so no prep is needed to cut the rest of the parts.

Turn on your router and hit the Start Cycle button as before. At this point there are more than 4400 instructions that need to be processed, which should take approximately six minutes to complete. Once done, shut down your router and vacuum the debris.

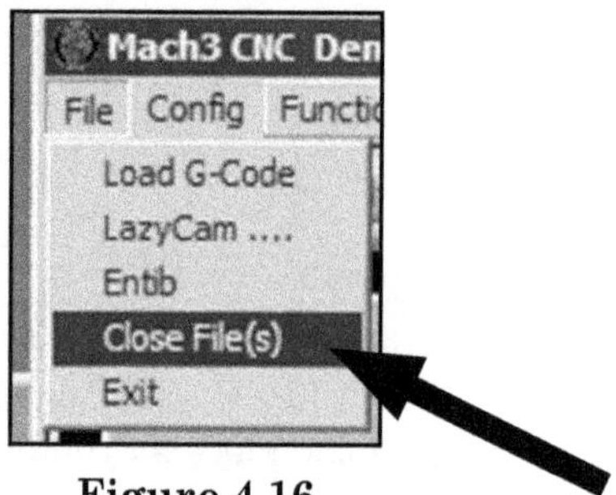

Figure 4.16

Step 9 - Cleanup the Parts

The quickest way to clean up the frayed edges on hardboard is by using a small palm sander like the one shown in Figure 4.17.

Remove the part from the CNC and hit each side with 120 grit sand paper. You don't need to do much; a few seconds on each side should be all that is necessary to smooth the edges.

Figure 4.17

The cleaned up parts should look like the ones shown in Figure 4.18. Detach them by pushing downward on the part to break the tabs, then remove the tabs with a pair of wire cutters.

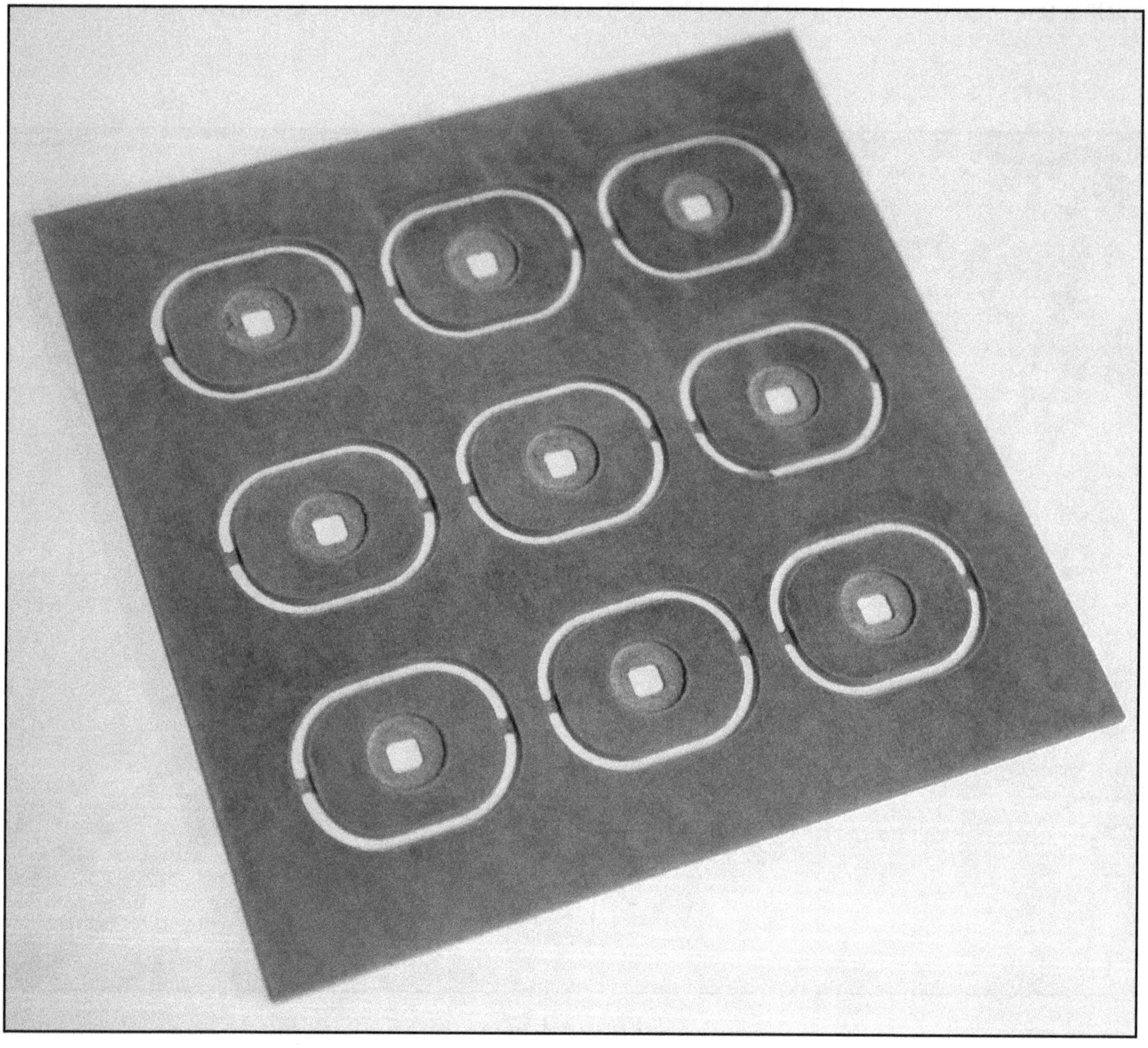

Figure 4.18

To assemble the parts, take any 1/4-20 carriage bolt and slip it through the bushing. Seat the head into the pocket, as shown in Figure 4.19. The corners of the square holes are a little tight so they should easily hold each bolt in place.

Figure 4.19

Conclusion

This chapter has presented you with the bare bones basics on using your CNC. Operational upgrades such as homing switches and probes will make setup quicker and more accurate, but it is important that you understand the basics before proceeding. Feel free to make a few more sets of clamp bushings. The G-code file called "All Clamp Bushings" contains instructions for making all nine bushings in a single file and is available on the Kronos Robotics web site.

Figure 4.20

Chapter 5

Dragon Cable

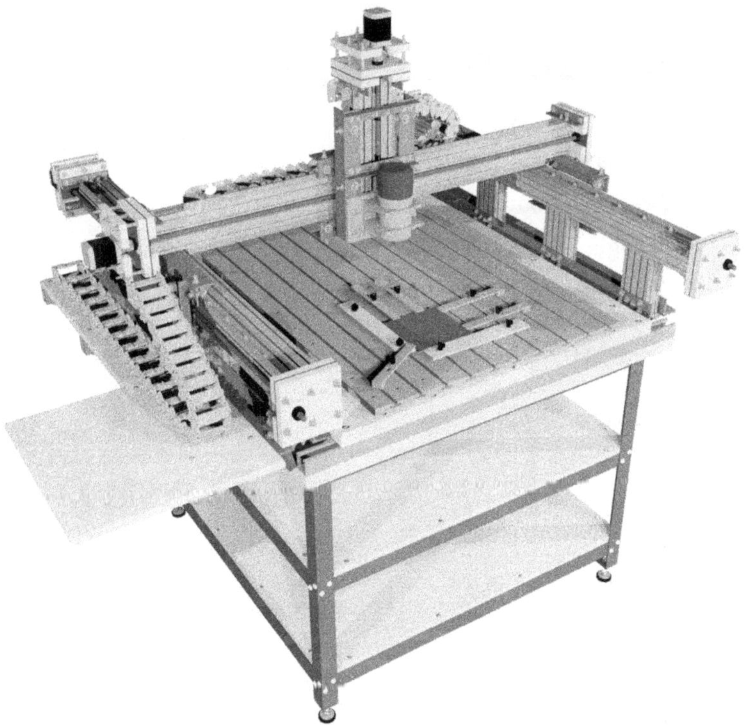

The dragon cable is an essential upgrade to the KRMx01 CNC. It will allow you the ability to run all your cables without hanging them. Later when you add future upgrades it will provide a way to keep all your cables out of harms way.

Tools Needed For This Chapter

- 2, 7/16 Wrenches or sockets
- 2, 9/16 Wrenches or sockets
- Portable drill
- Palm sander or sanding sponge
- Freud 1/16" router bit
- Drill press
- 5/16" and 1/4" Drill bits

Components Needed For This Chapter

- 3, 24" x 48" x 1/8 Hardboard
- 1, 48" x 4-1/4" x 3/4" Particle board (see text)
- 1/2" 4 hole 90 degree angle bracket (strut bracket)
- 2, 1/4-20 x 1-1/2" Carriage bolts
- 2, 1/4-20 Hex nuts
- 6, 1/4" Washers
- 2, 1/4" Split lock washers
- 2, 3/8-16 x 1-1/2" Hex bolts
- 1, 3/8-16 x 3" Hex bolts
- 29, 3/8" Washers
- 4, 3/8" Split lock washers
- 5, 3/8-16 Hex nuts
- 260, #6 x 1/4 Machine screws
- 135, #6 x 3/8 Machine screws

The 4 hole 90 degree bracket can be found in the electrical section at most home centers. They are used to attach square conduit.

Time Estimates

Cost: $32

Time: Weekend

Prerequisites

Cut the 3, 24" x 48" x 1/8" panels into 54, 7-7/8" x 7-7/8" squares, as shown in Figure 5.1. You actually only need 40 panels, 39 for the parts, and one for the waster board.

Figure 5.1

To complete this project you will need the G-code files located on the Kronos Robotics web site. You can download them here:

http://www.kronosrobotics.com/krmx01/da69781_9386_87112/

You will need a Freud 1/16" router bit shown in Figure 5.2. The part number for this bit is #04-096. You can purchase them from Amazon.

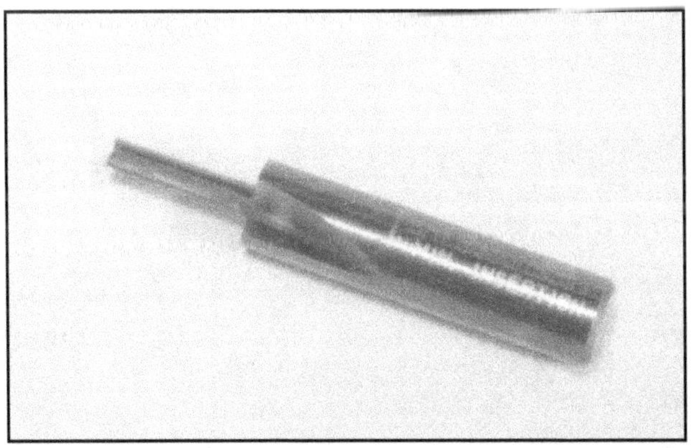

Figure 5.2

The X dragon cable support is a piece of 48" x 4-1/4" x 3/4" stock. It can be made from any material available. It can also be made from solid wood if you prefer as its only job is to support the dragon cable. You will need a drill press to drill the holes needed to mount it to the X-beam.

Step 1 - Mount Stock to Table

Take one of the 7-7/8" x 7-7/8" x 1/8" pieces of hardboard and stack it on top of a second 7-7/8" x 7-7/8" x 1/8 piece. The bottom piece will become the waster board. Place them on your clamp table against the two fences, as shown in Figure 5.3. Use two clamps on each of the exposed sides, as shown in Figure 5.3. The clamp in the front left corner should be slightly offset, exposing the corner as shown.

Figure 5.3

Tip

Sand the tip of the corner clamp to a point. This will make it easier to expose the corner of your stock.

Step 2 - Position the Router

Power up the PC and load Mach 3. Apply Power to your 48V power supply.

Remember you need to make Mach 3 active by hitting reset.

Place the Freud 1/16" router bit in the router. About 1" of the bit should be exposed.

Set the jog speed to 10%, and using the arrow keys move the tip of the router bit to the corner of the work piece as shown in Figure 5.4. Use the page up and page down keys to set the bit exactly on the surface of the top part. You want the corner tip of the stock to be dead center of the router bit.

As in the last chapter hit the "Ref All Home" button and then zero each of the axis by hitting the X, Y, Z Zero buttons.

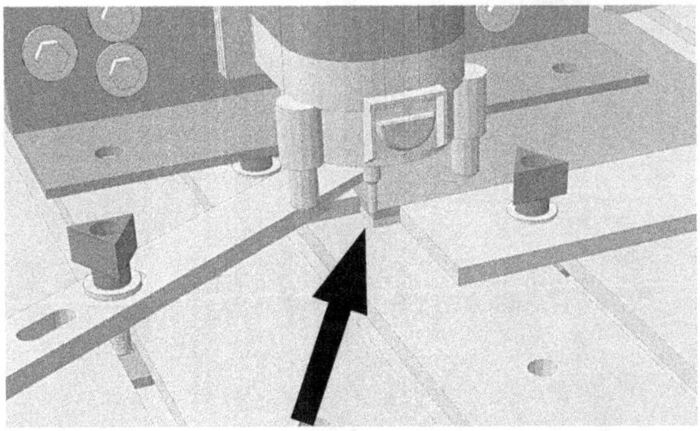

Figure 5.4

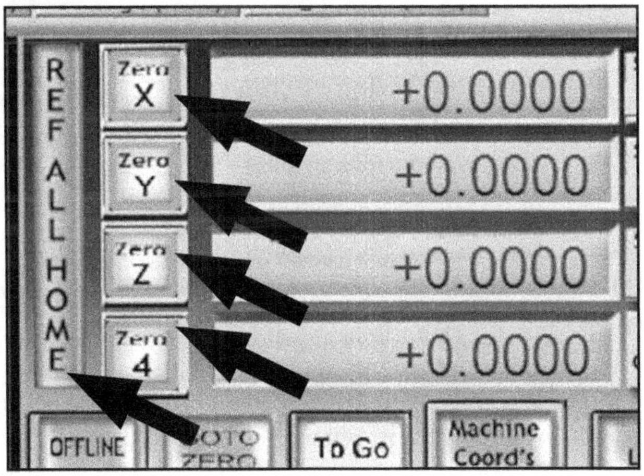

Figure 5.5

Tip

The zero sequence mentioned in the previous paragraph is only needed if your CNC machine does not have homing switches and does not automatically home on your two fences.

Step 3 - Load the Full Link G-code

Use the Load G-code option in the File Menu to load the “Dragon Cable Full Link.txt” file.

The small display window on the right will display the part layout shown in Figure 5.6.

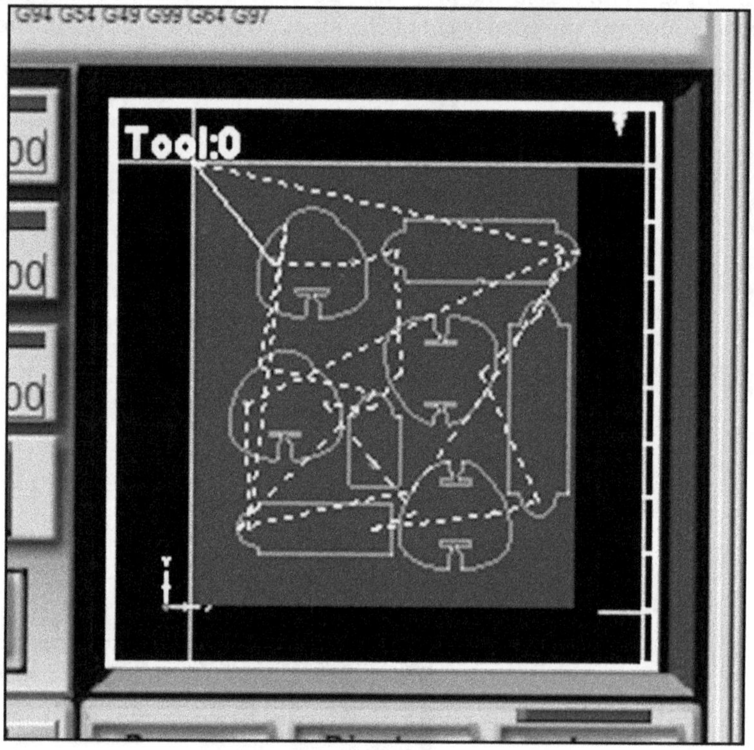

Figure 5.6

Tip

The G-code for this file is available on the Kronos Robotics web site at www.kronosrobotics.com.

You can download the file here:

http://www.kronosrobotics.com/krmx01/da69781_9386_87112/

Step 4 - Make the Part

Raise your Router so that the bit is a couple inches above the table.

Set the speed of the router to 4 and start the router.

Hit the Cycle Start Button shown in Figure 5.7.

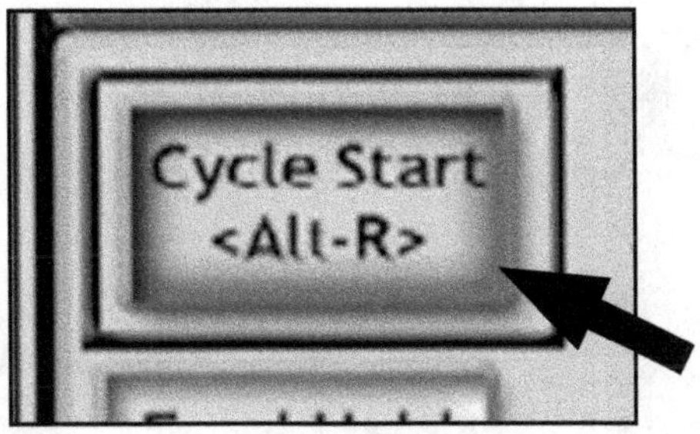

Figure 5.7

The finished part should look like the one shown in Figure 5.8. Use a palm sander with 120 grit paper to clean the burrs from the cut edges. Do this on the complete piece before removing the parts. Note that you can also use a fine sanding sponge.

You are going to build the X axis cable first so you will need 15 complete sets cut.

When a part is complete, remove it, vacuum the dust and clamp the next part in place. Then hit the Cycle Start button again. Each part takes about 10 minutes to complete. You will use the same waster board under all the parts cut in this chapter.

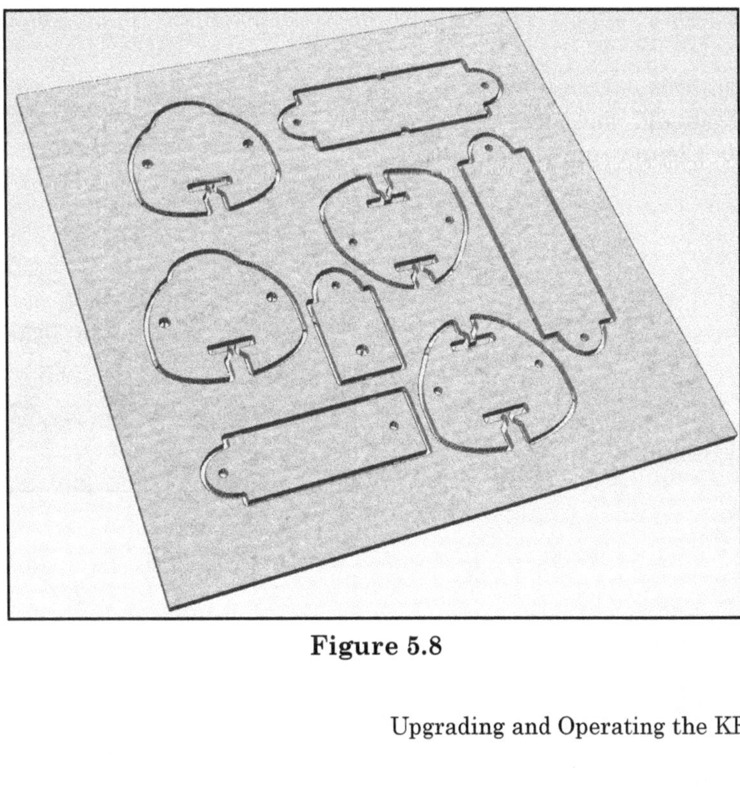

Figure 5.8

Step 5 - Make the Tail

Once the 15 links for the X cable have been cut, close the file. Load the file called "Dragon Cable tail. txt". The part tool paths will show up in the display as shown in Figure 5.9.

Clamp a piece of stock to your clamp table, start the router, then hit the Start Cycle button to make the part.

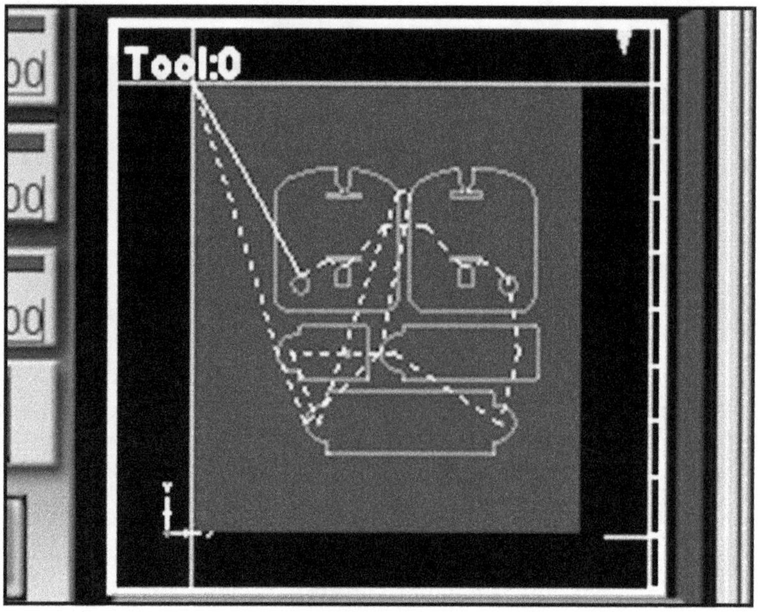

Figure 5.9

The finished part should look like the one shown in Figure 5.10. Clean it up with a sander and set it aside.

Figure 5.10

Step 6 - Make the Head

Close the current file and load the file called "Dragon Cable head.txt". As before, the part tool paths will show up in the display as shown in Figure 5.11.

Clamp a piece of stock to your clamp table, start the router, then hit the Start Cycle button to make the part.

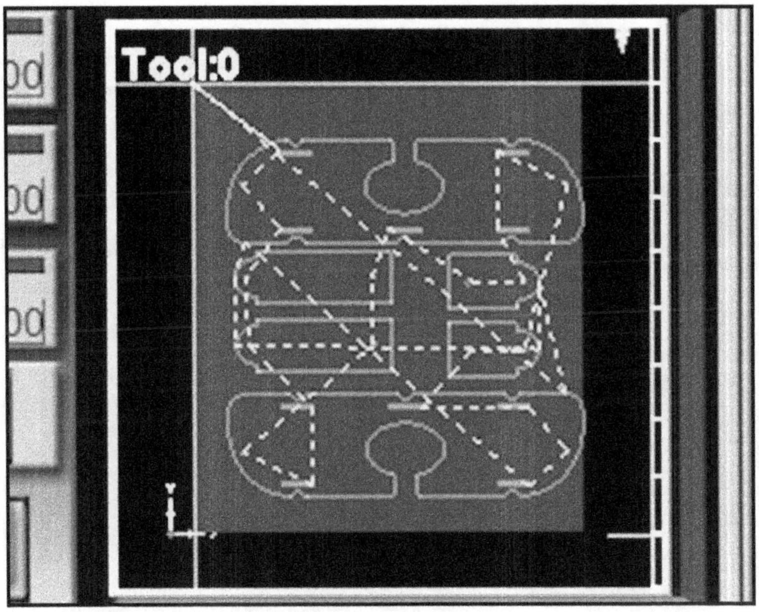

Figure 5.11

The finished part should look like the one shown in Figure 5.12. Clean it up with a sander and set it aside.

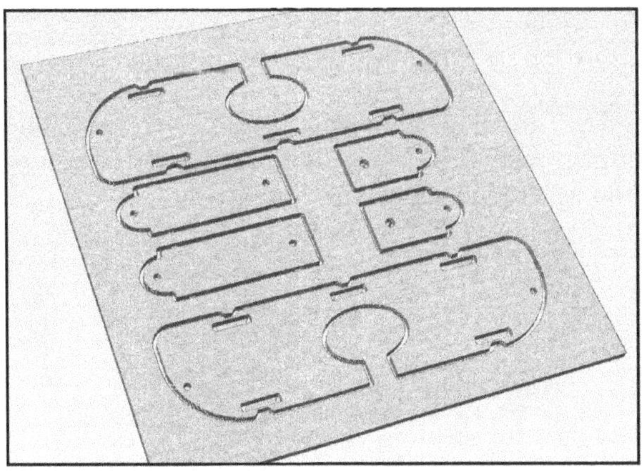

Figure 5.12

Step 7 - Make the Head Base

Load the file called "Dragon Cable head base.txt". Again, the part tool paths will be displayed as shown in Figure 5.13.

Clamp a piece of stock to your clamp table, start the router, then hit the Start Cycle button to make the part.

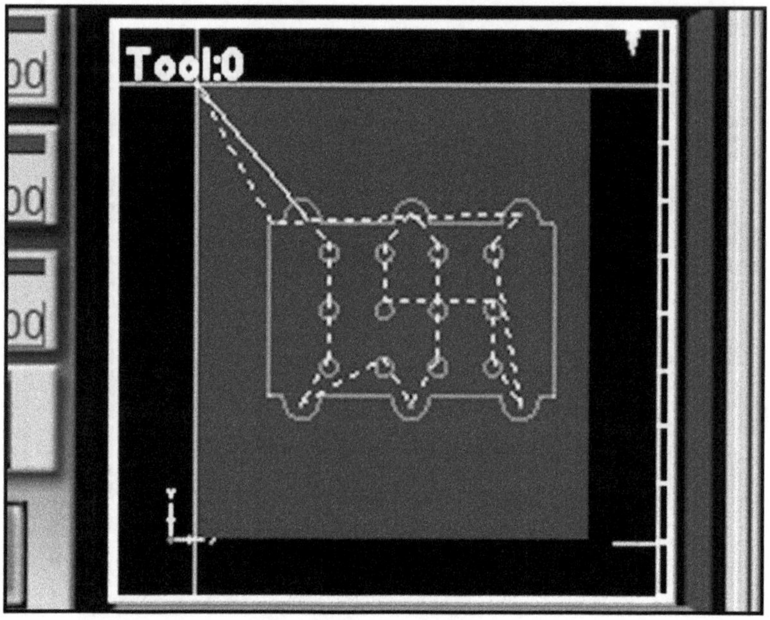

Figure 5.13

The finished part should look like the one shown in Figure 5.14. Clean it up with a sander and set it aside.

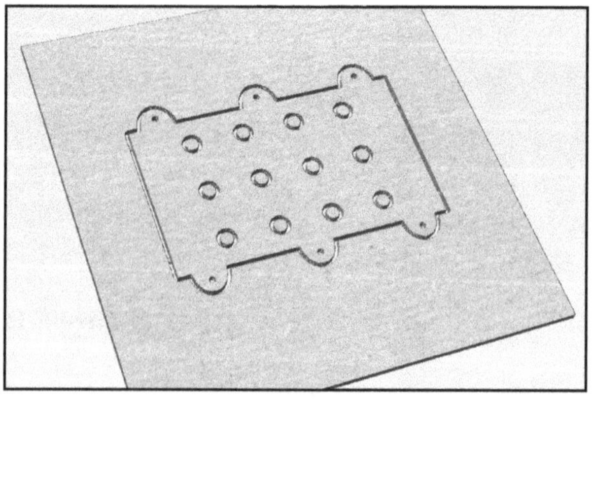

Figure 5.14

Step 8 - Assemble the Links

Punch out all the parts in each of the 15 sets of full links. This will yield several parts. It is recommended that you separate all the parts into similar groups.

Tip

Always punch the parts out from the top (cut side). This will ensure the tabs don't chip the part. Instead they will break off part of the waist stock.

There is an exception to this rule. If you are punching out a cutout from the interior of the finished part, you push the part out from the bottom.

There are two link types. An inside link and an outside link. The inside link, like the one shown in Figure 5.15, has sides that can be identified by the lack of a slot at the top. The lower insert can be identified by a small round indent on the sides.

To assemble the inside link insert the round tab in the end of the lower insert into the slot and lock it in place with a 1/4" #6 machine screw.

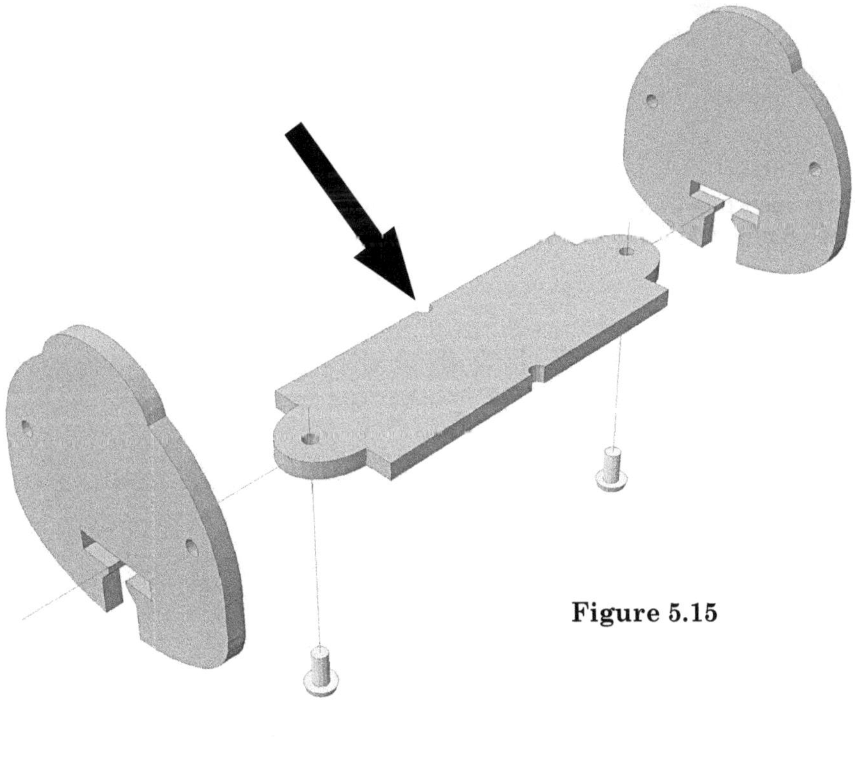

Figure 5.15

The outside link like the one shown in Figure 5.16, has sides that have slots at both the top and bottom. The lower insert does not have the small indents. It is also slightly longer.

Assemble the outside link the same way as the inside link.

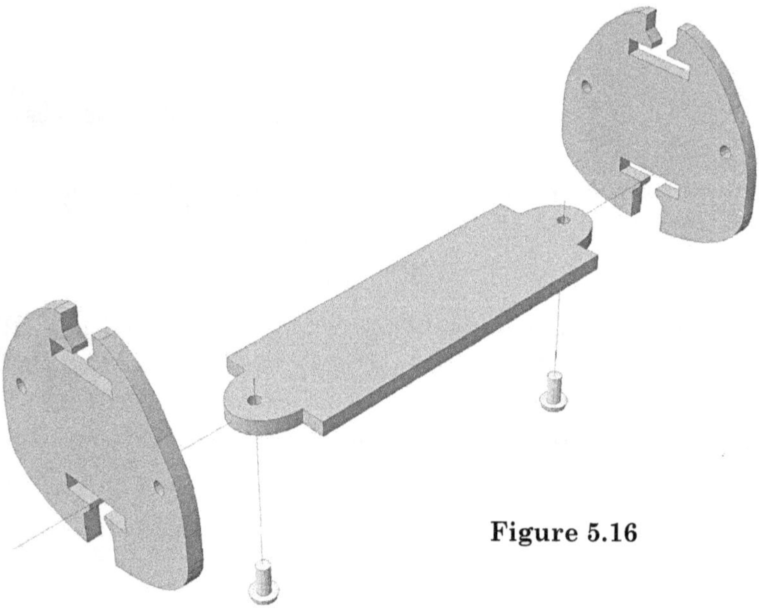

Figure 5.16

The outside link also has four 3/8" #6 machine screws inserted into the side pieces, as shown in Figure 5.17. These will act as hinge points on the link.

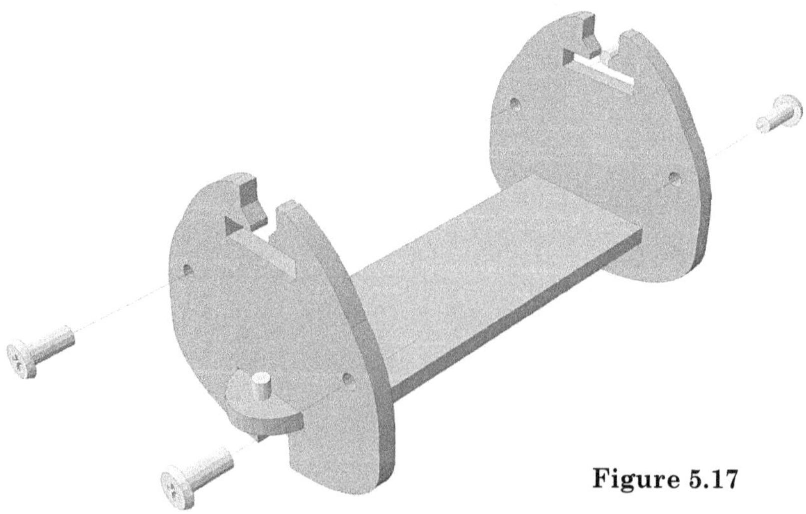

Figure 5.17

Tip

Throughout this document, the term full link and link is used. When the term full link is used it referring to both a inside and outside link as a whole.

Insert two of the 3/8" machine screws all the way on one side of the link. Insert the other two 3/8" machines screws so that only about 1/16" protrudes through, as shown in Figure 5.18. Call this the short screw. You will tighten it later once the two links are attached.

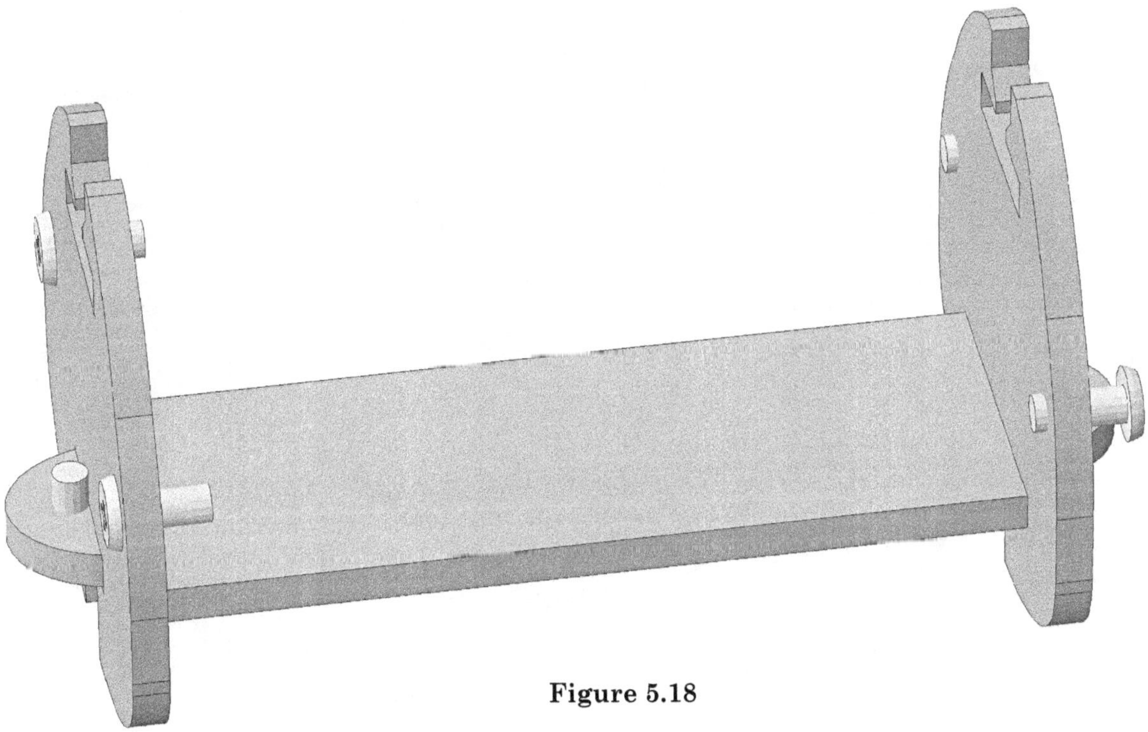

Figure 5.18

To connect the two links take the outside link and slip the end with the long screw into the hole of the inside link. Then snap the shorter screw into the mating hole on the inside link.

Once in place, finish tightening the short screw until its fully seated. The finished link should like the one shown in Figure 5.19.

Finish the process by alternating inside and outside links, until all 15 full links are connected. Don't attach the last outside link, it's an extra.

When completed you should have 29 links connected together with an inside link at both ends of the dragon cable.

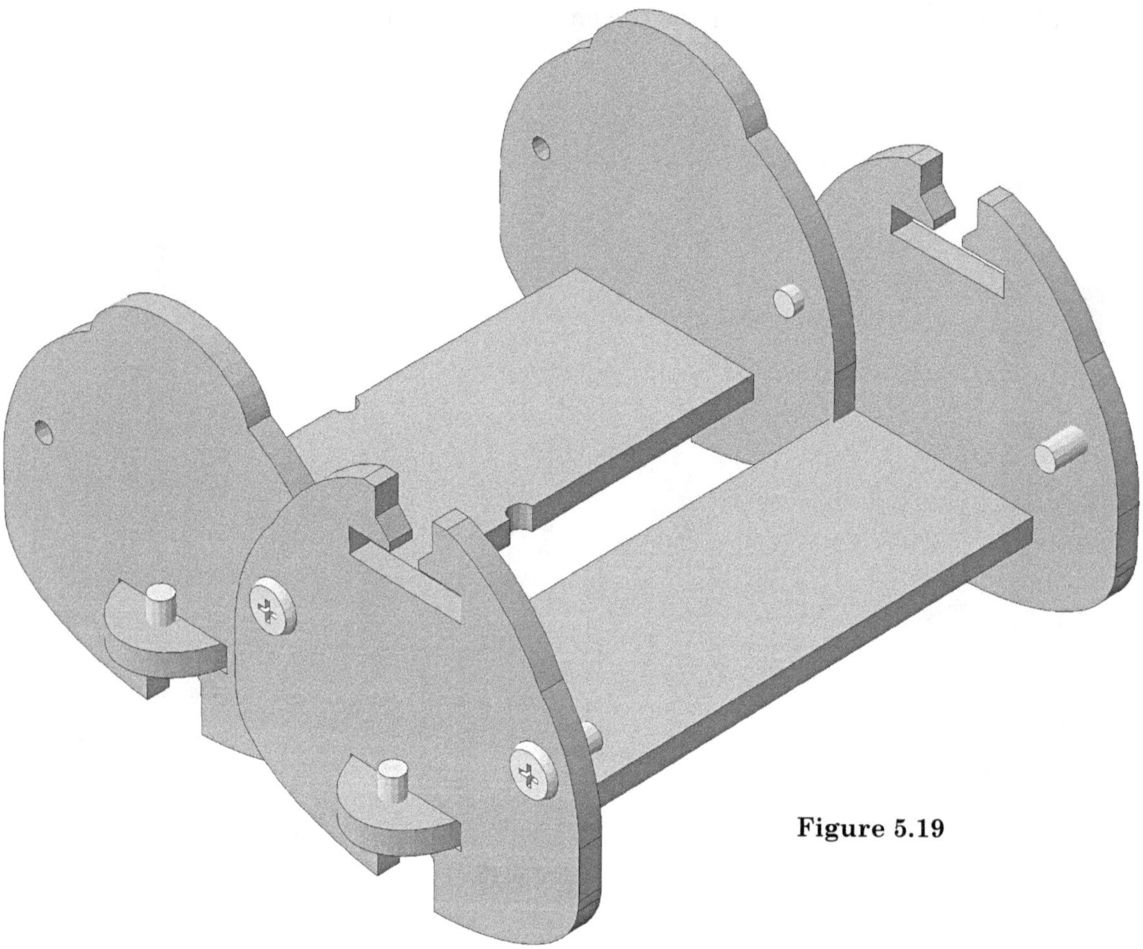

Figure 5.19

Step 9 - Assemble the Head

Punch out all the dragon cable head parts from the two pieces of stock you cut earlier.

The head is assembled in much the same way as the outside links. It's just a lot bigger.

Insert the three round tabs on the head base into the three lower slots on the head sides as shown in Figure 5.20. Lock them in place with a 1/4" #6 machine screw.

As before insert a 3/8" #6 machine screw all the way into the rear end of one of the sides as shown here. The opposite side screw should only be inserted part way as before.

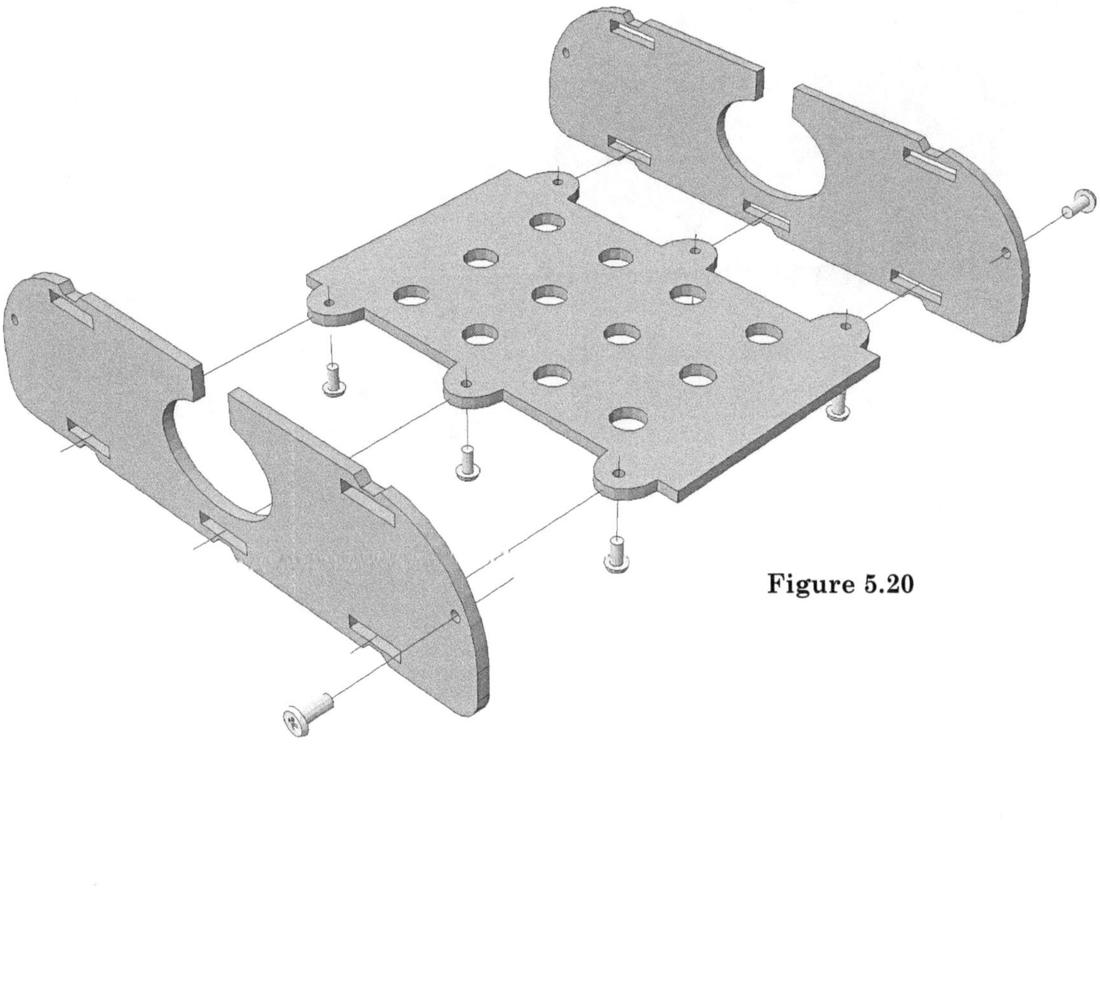

Figure 5.20

Attach the head to the end of the dragon cable just like you did with the outside link. Tighten the short screw once in place, as shown in Figure 5.21.

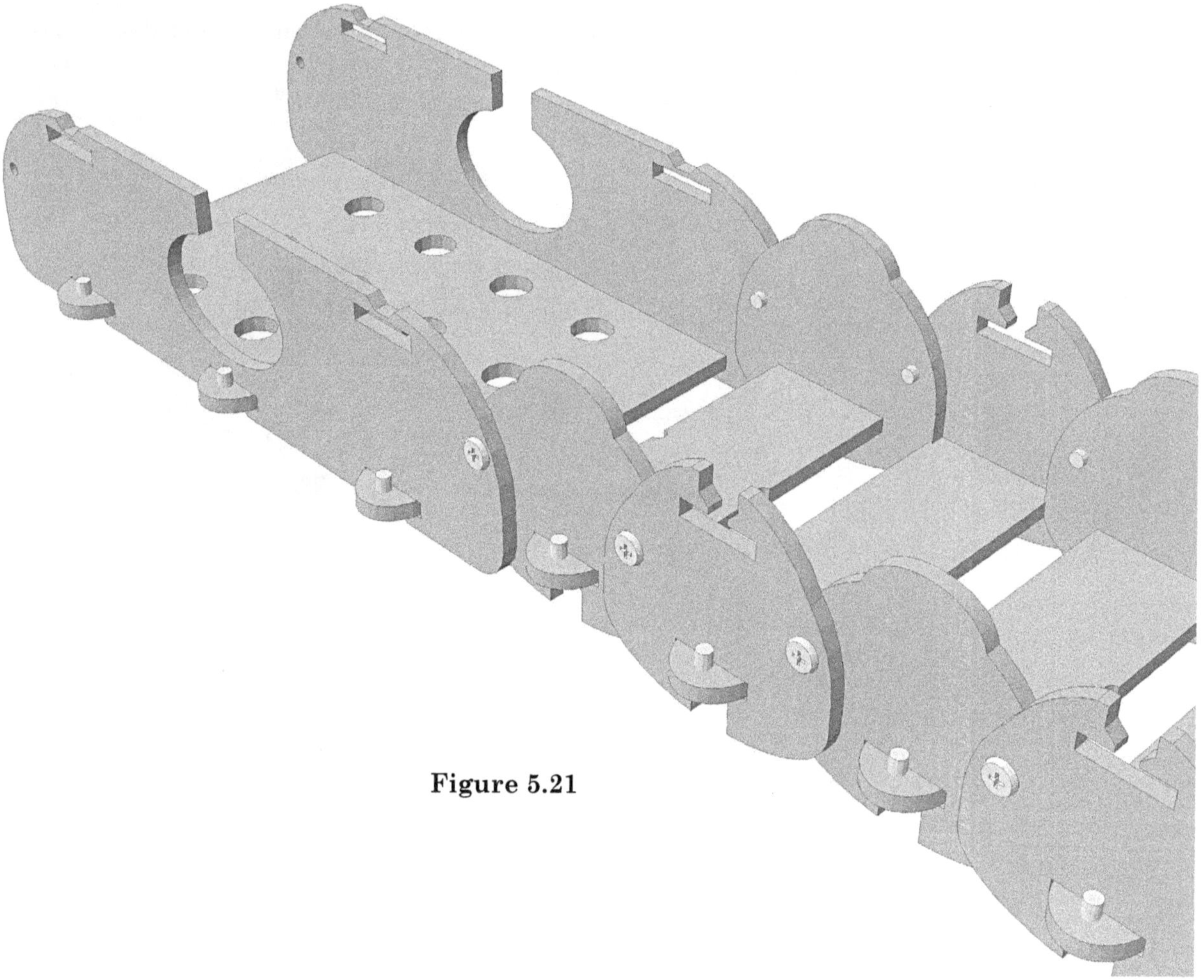

Figure 5.21

Step 10 - Assemble the Tail

Punch out all the dragon cable tail parts from the piece of stock you cut earlier.

To assemble the inside link insert the round tab in the end of the lower insert into the lower slot and lock it in place with a 1/4" #6 machine screw, as shown in Figure 5.22.

Insert a 3/8" #6 machine screw all the way into the front end of the left side as shown here. The right side screw should only be inserted part way as before.

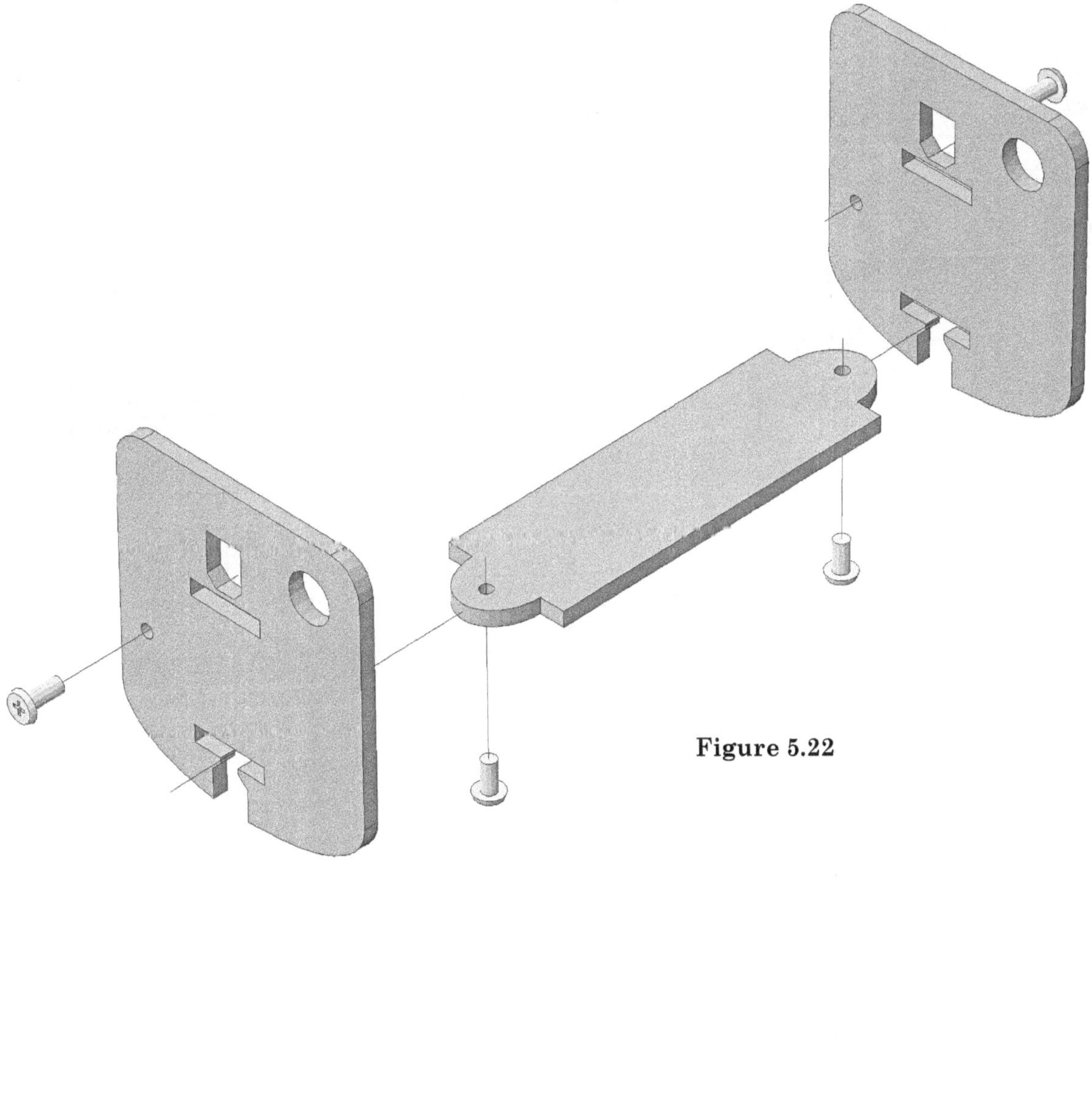

Figure 5.22

Attach the head to the end of the dragon cable just like you did with the outside link. Tighten the short screw once in place, as shown in Figure 5.23.

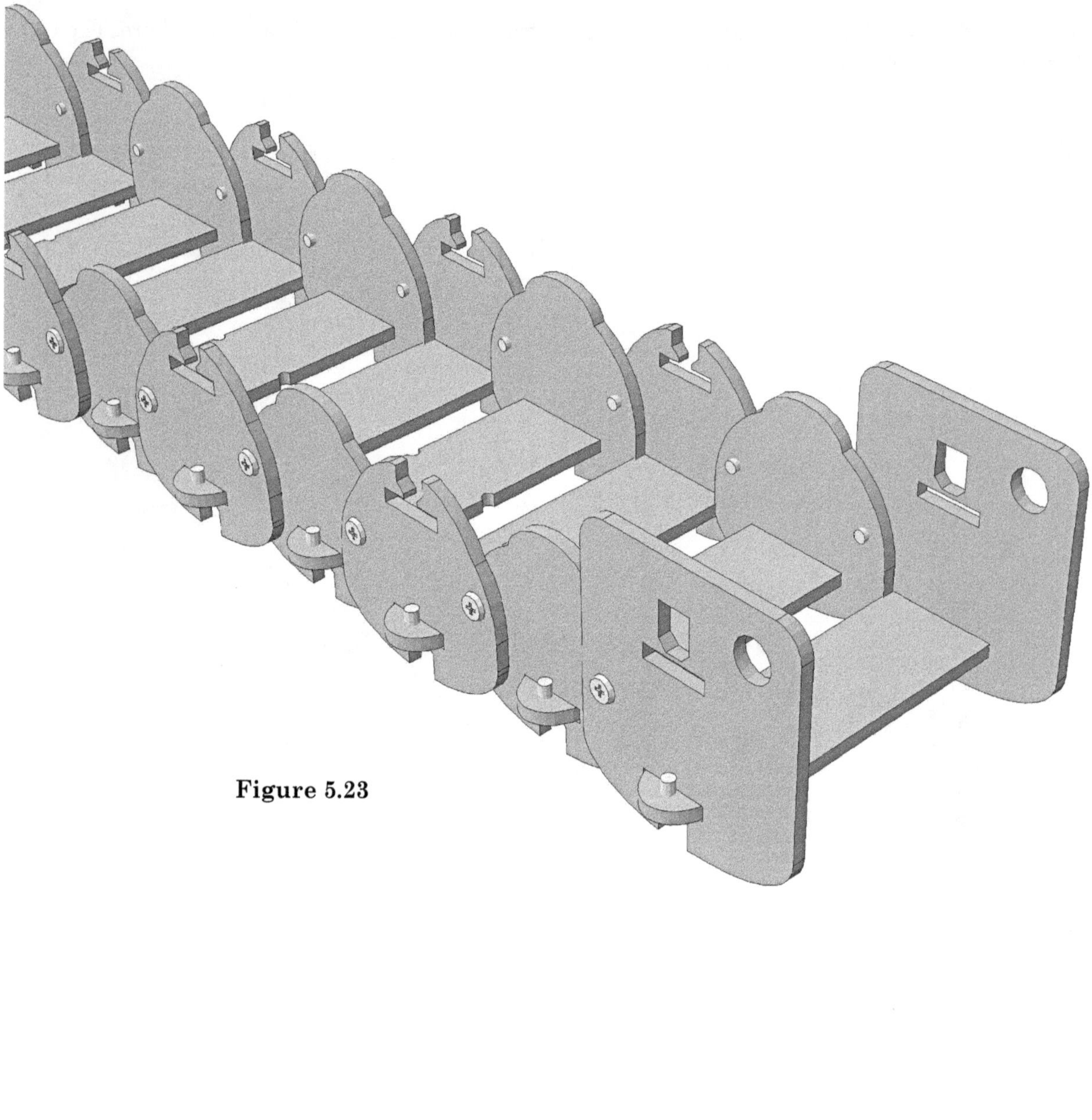

Figure 5.23

Step 11 - Add Upper Inserts

The upper inserts consists of two components. A shorter insert and a longer insert. They are attached in the same way as the lower inserts, as shown in Figure 5.24.

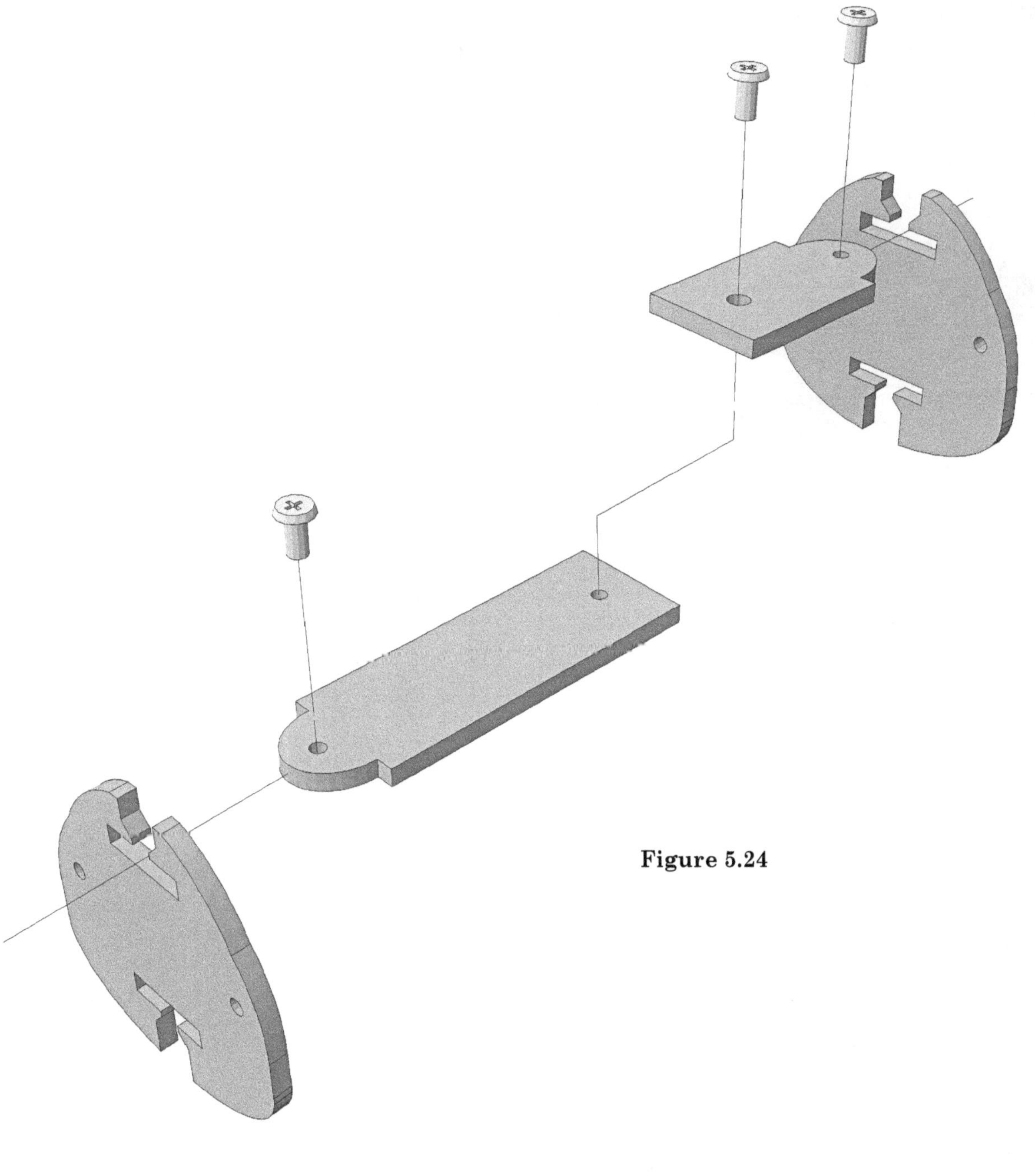

Figure 5.24

Take the longer upper inserts and insert them into the upper slots of the dragon cable as shown here. Secure in place with a 1/4" #6 machine screw.

Be sure to add the insert to the head and tail as shown in Figures 5.25 and 5.26. Observe the orientation of the cable.

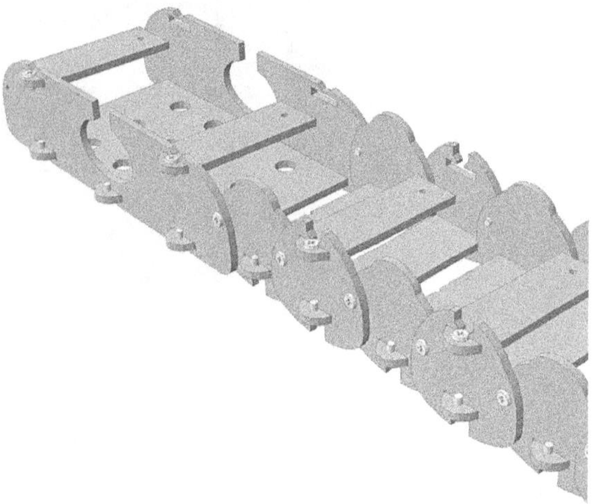

Figure 5.25

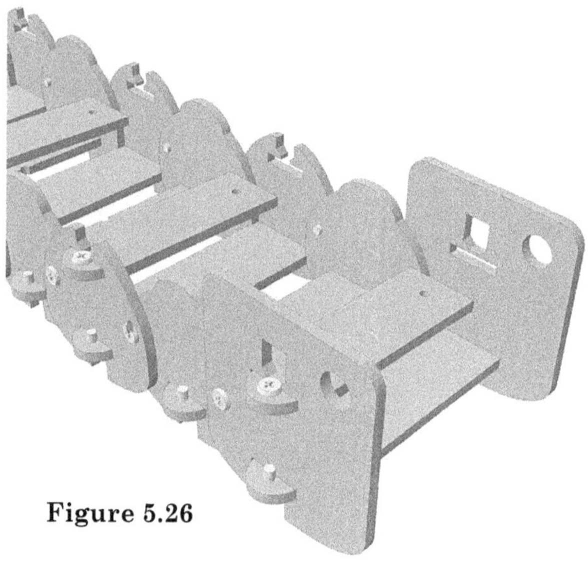

Figure 5.26

Take the shorter upper inserts and insert them into the upper slots of the dragon cable as shown in Figures 5.27 and 5.28. Secure in place with a 1/4" #6 machine screw.

Don't connect the two upper supports at this time. You will do that once your cables are installed.

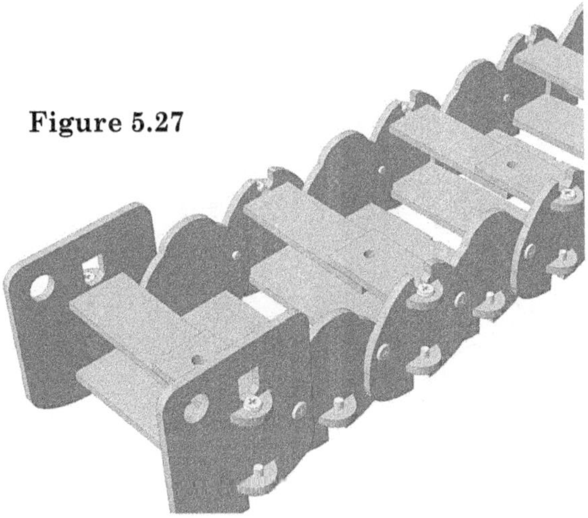

Figure 5.27

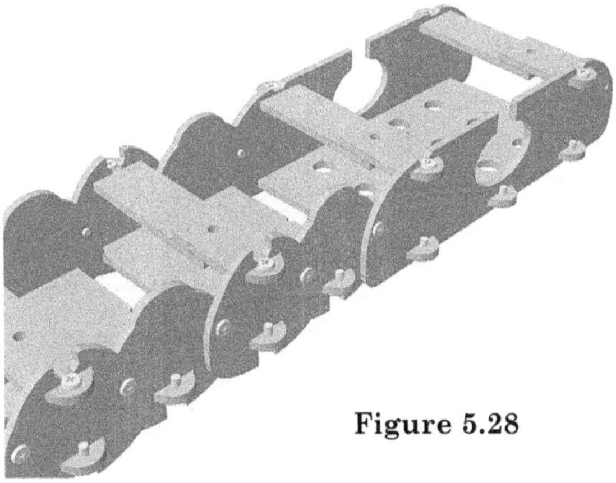

Figure 5.28

If you painted you dragon cable now would be a good time to touch up the paint job.

Step 12 - Create X-beam Cable Support

The X-beam dragon cable support can be made from any material as long as its 3/4" thick. The support is 48" long by 4-1/4" wide. Use a drill press to drill the holes shown in Figures 5.29 and 5.30. The 7 holes in the edge of the support are 5/16" in diameter. The single hole on the end face of the support is 1/4" in diameter.

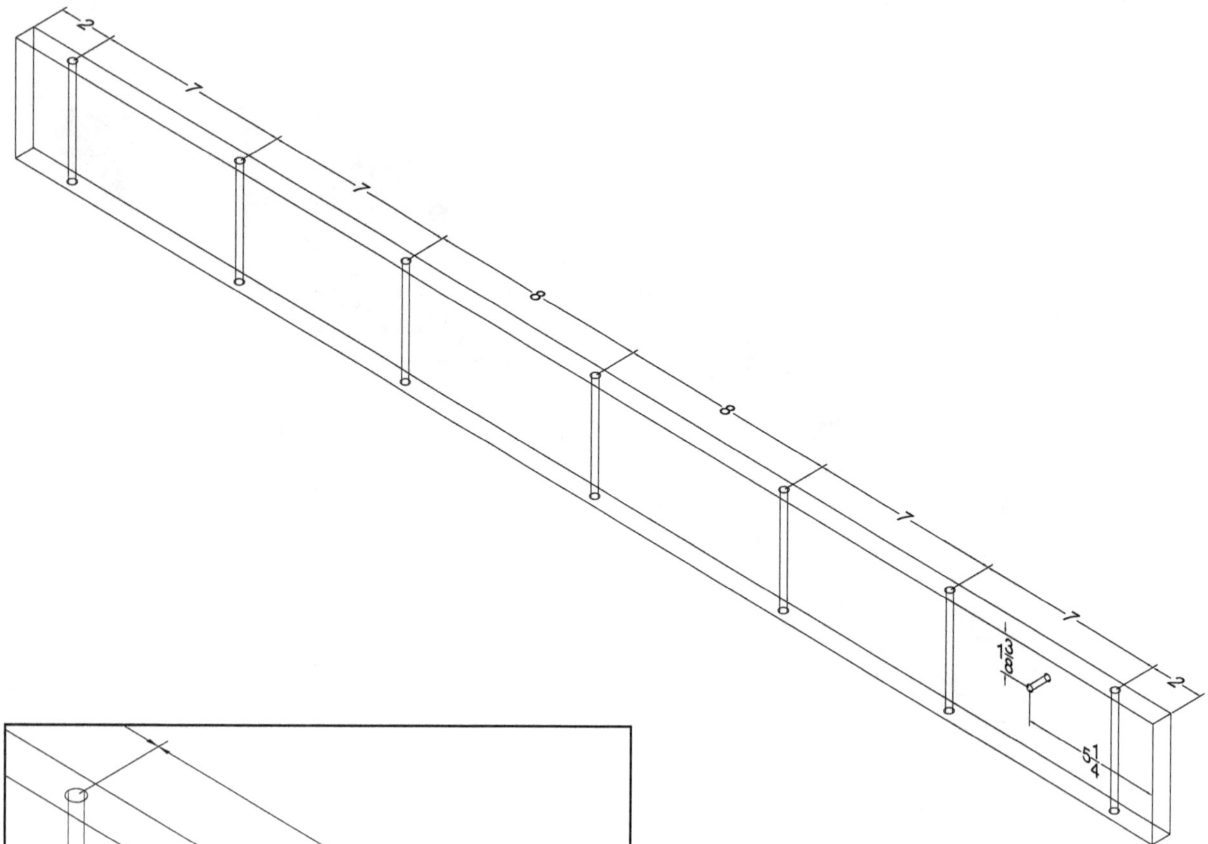

Figure 5.29

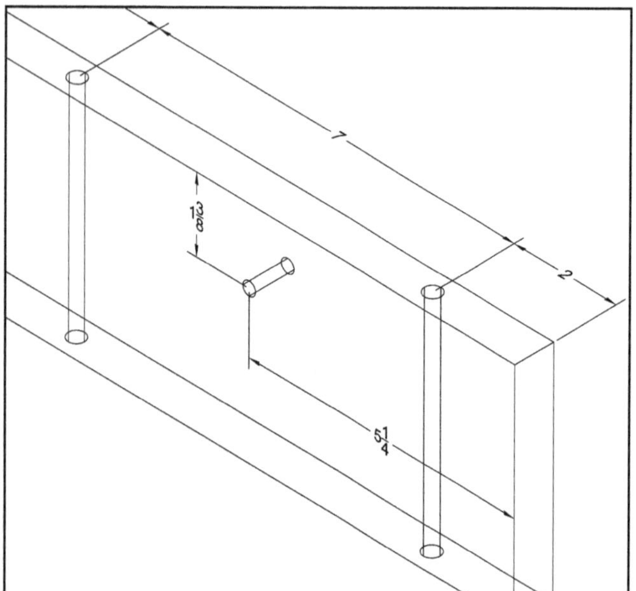

Figure 5.30

Step 13 - Mount X-beam Cable Support

The KRMx01 assembly instructions had you insert seven carriage bolts when assembling the X-beam. Space the seven bolts so that they mate with the 7 holes in the support. The support should be centered on the X-beam. That should place the edge of the support about 4" from each edge of the beam.

Place a 1/4" washer, lock washer, and 1/4" hex nut on each of the exposed bolts and tighten, as shown in Figure 5.31.

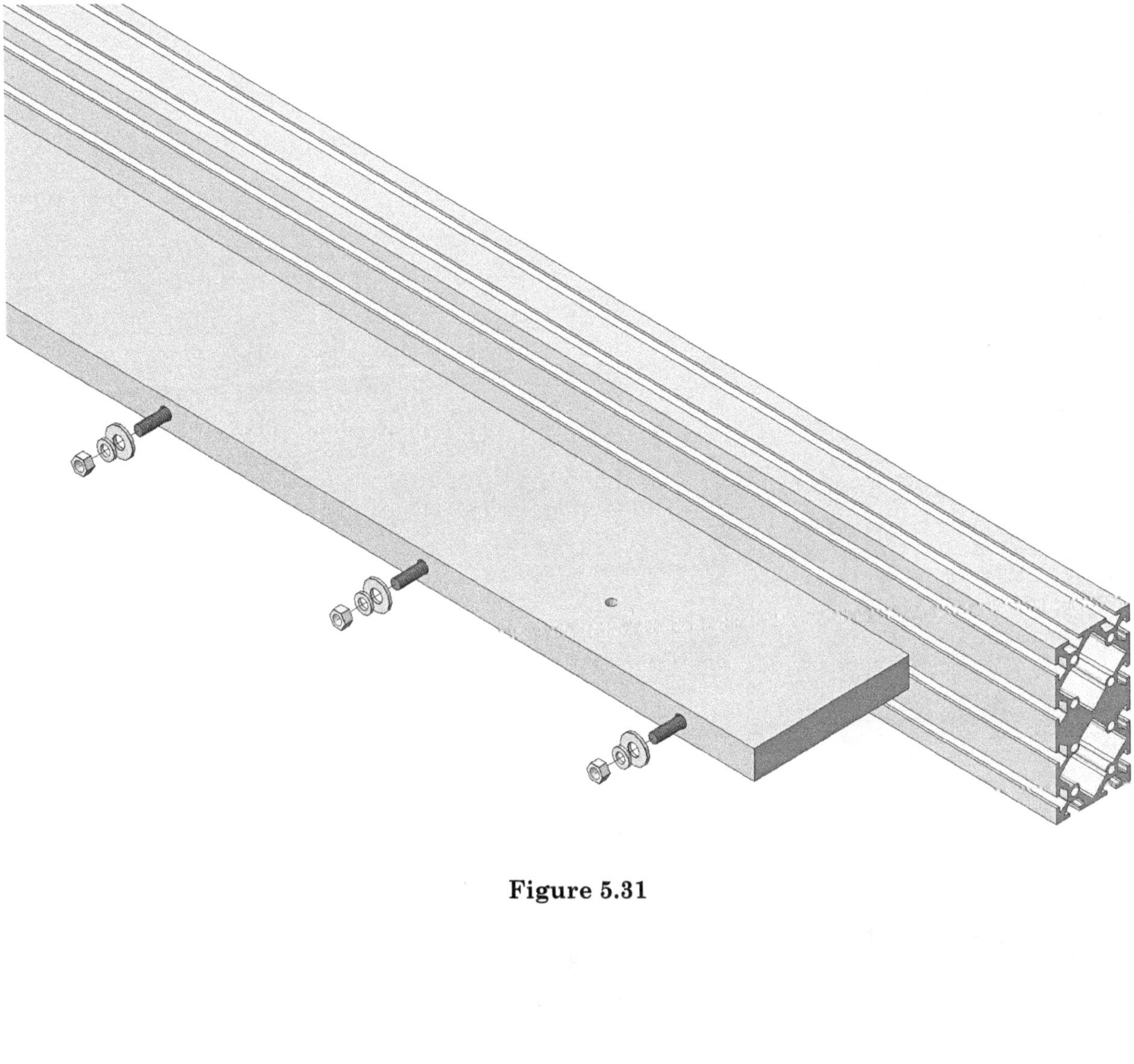

Figure 5.31

Step 14 - Attach X Dragon Cable Mounting Hardware

On the right side (from rear) of the X support ,is a 1/4" hole. Insert a 1/4-20 x 1-1/2" carriage bolt through the bottom of the hole. Add two 1/4" washers as in Figure 5.32.

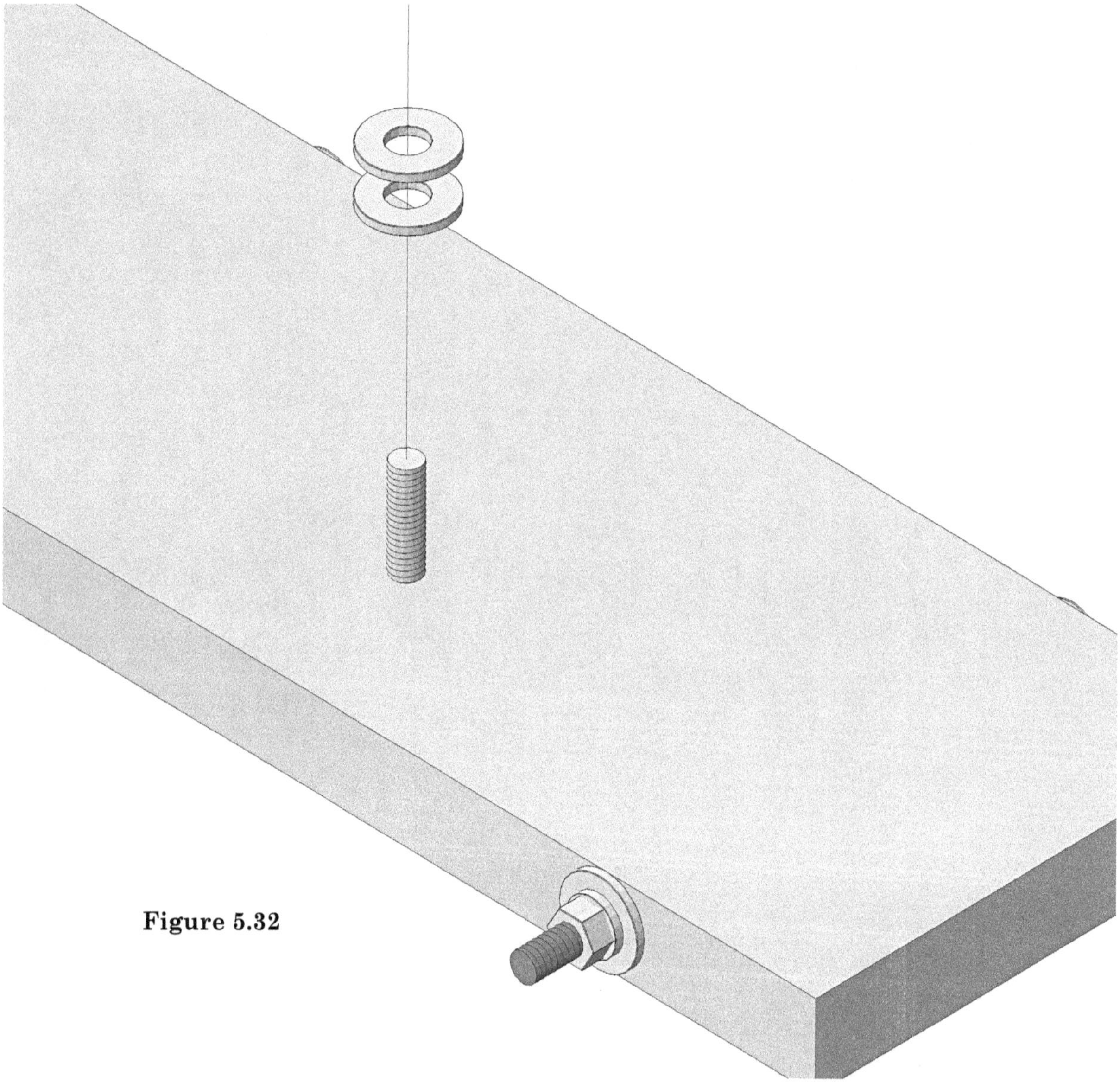

Figure 5.32

Take the right angle strut bracket and place the short edge on the top of the X-carriage, mating the rear hole to the 3/8" hole in the center of the carriage.

Attach with a 3/8-16 x 1-1/2" hex bolt as shown in Figure 5.33. Add a washer, lock washer and hex nut and tighten.

Take a 3/8-16 x 1/2" bolt and add a washer and insert into the upper hole in the bracket as shown in Figure 5.33. Add a washer, lock washer and hex nut and tighten. Add an additional washer after the nut.

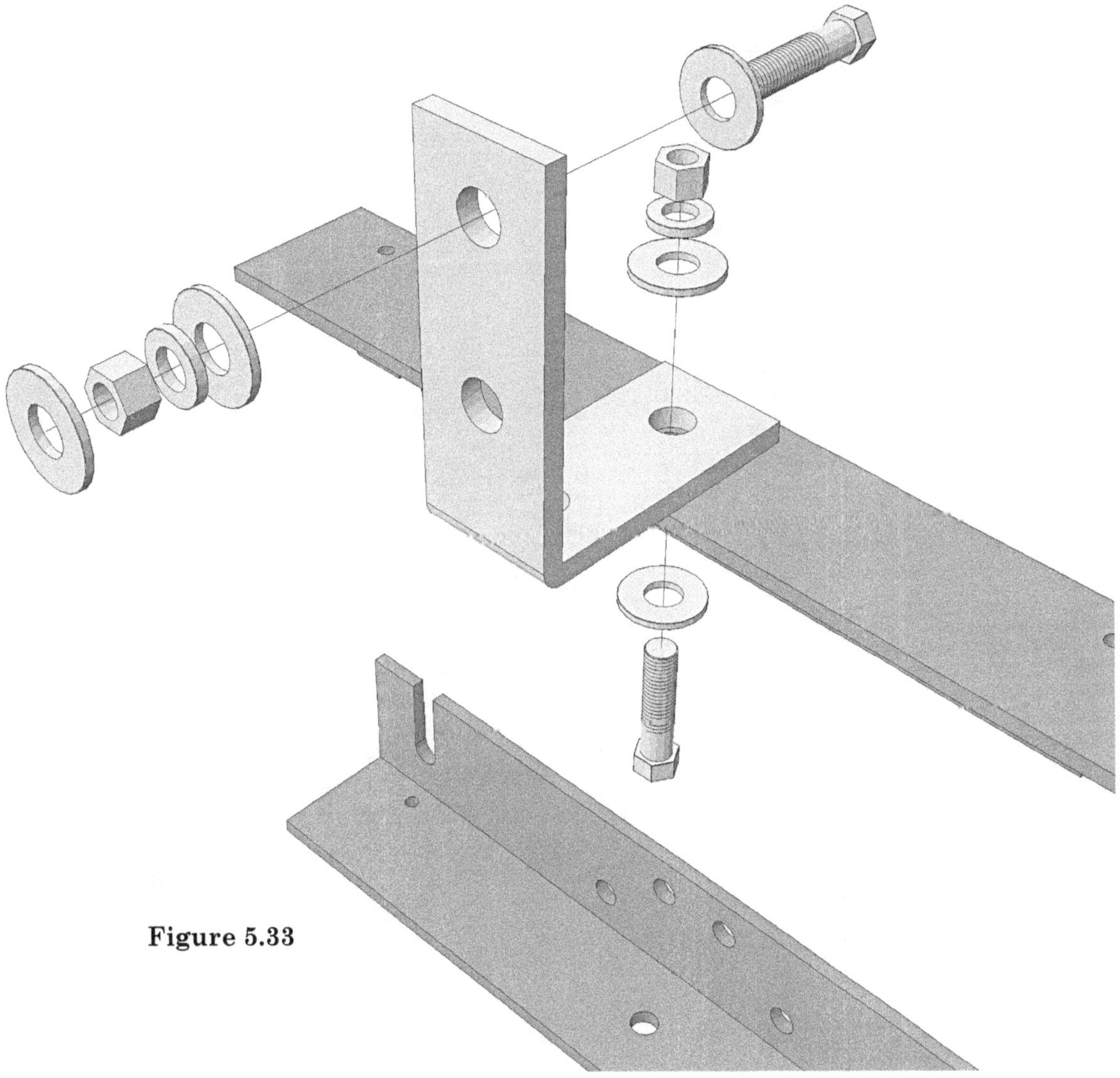

Figure 5.33

Place the head of the dragon cable over the carriage bolt on the support, as shown in Figure 5.34. The bolt should be protruding through the center hole, second from the end.

Add a 1/4 washer, lock washer and 1/4" hex nut as shown. Finger tighten only.

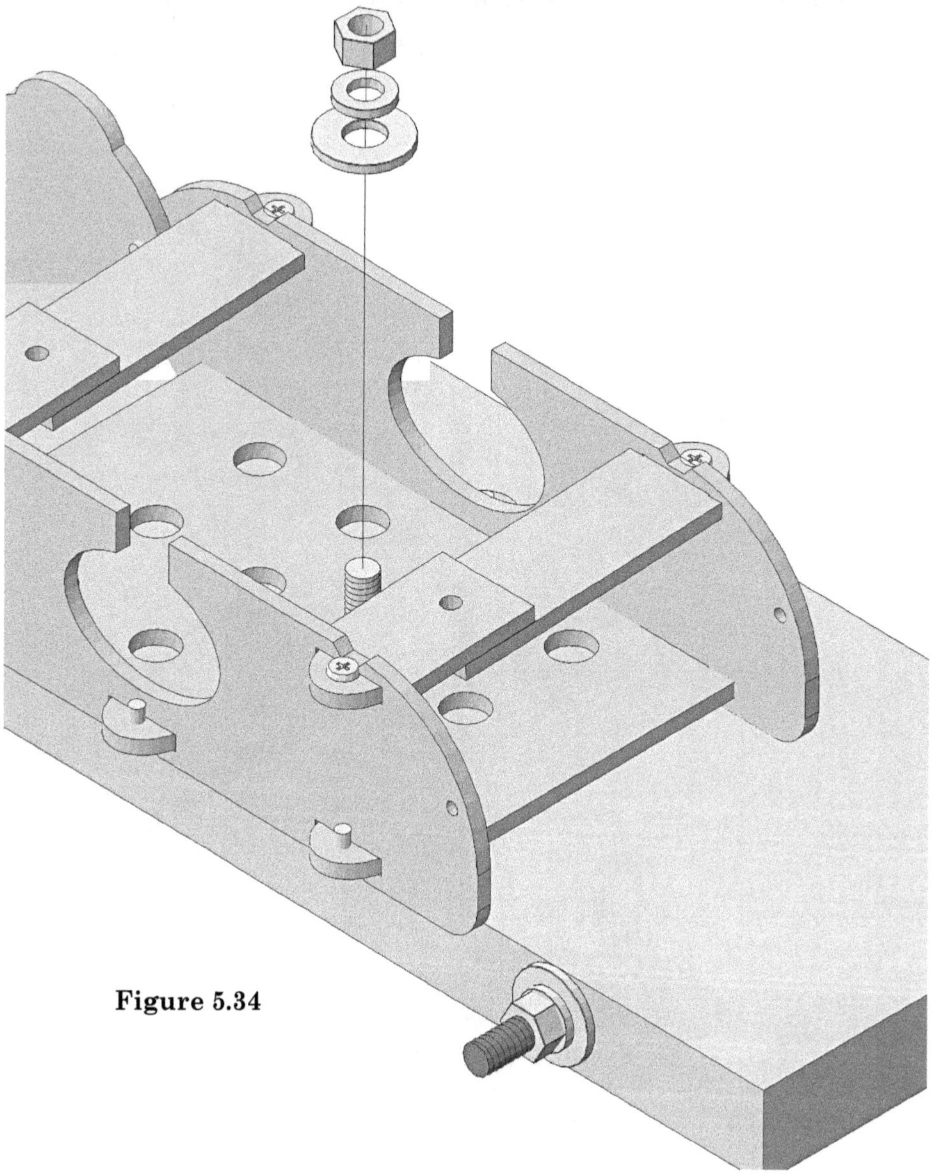

Figure 5.34

Wrap the tail around and place the 3/8 hole over the protruding 3/8" bolt as shown here in Figure 5.35.

Add a 3/8" washer lock washer, and 3/8" hex nut. Position the tail as shown and tighten the hex nut.

Go back and tighten the 1/4" hex nut on the head.

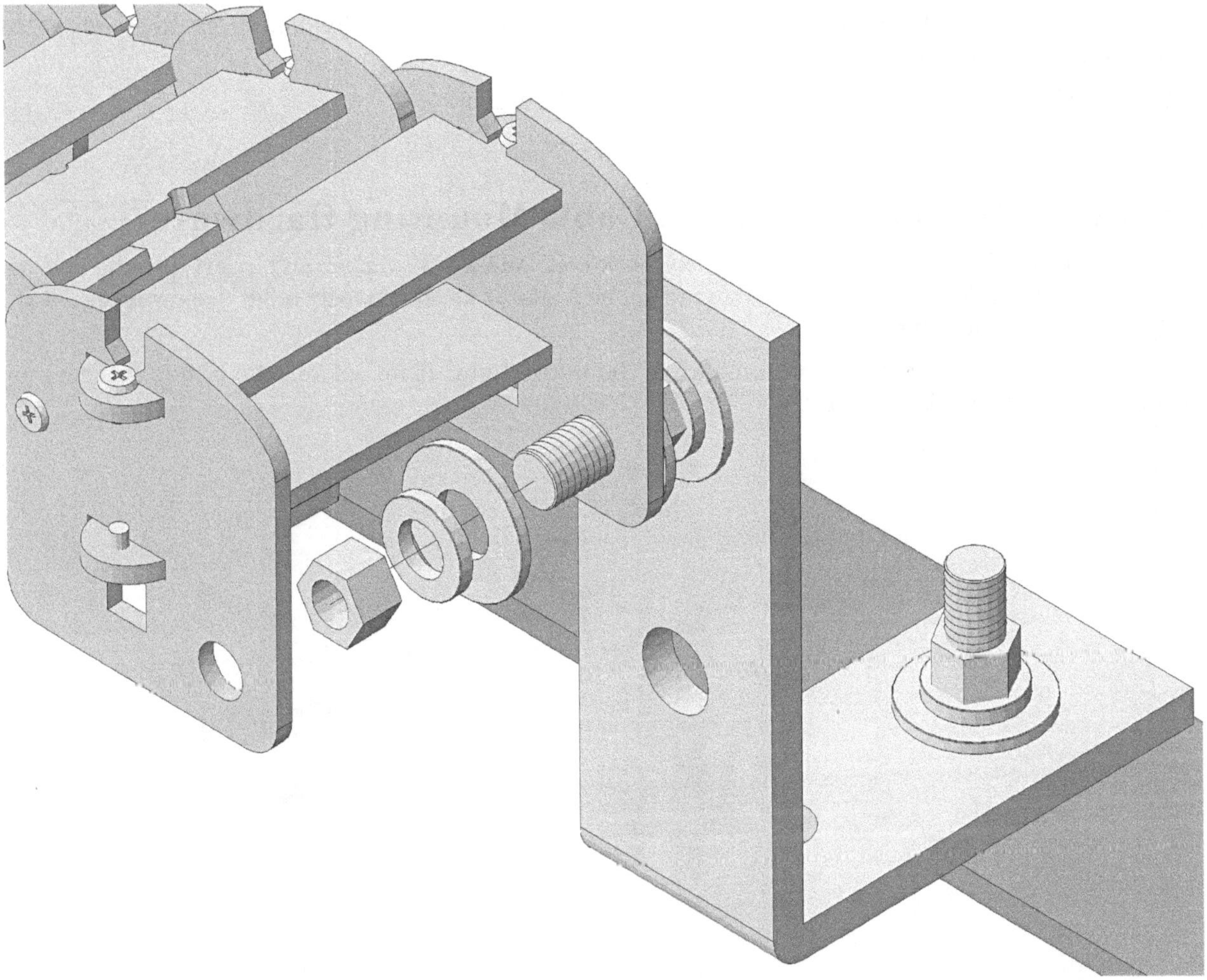

Figure 5.35

Step 15 - Make the Y Dragon Cable Parts

The Y dragon cable parts are made exactly the same way as the X cable. The only exception is that you need 18 full links instead of the 15 used on the X cable.

Step 16 - Assemble the Y Dragon Cable Parts

Assembly of the Y cable is the same as the X, however you will 35 individual links. Remember the two outside links are inner links only.

Add your head and tail the same way you did with the X cable.

Step 17 - Attach the Y Dragon Cable Mounting Hardware

On the left hand Y-carriage take a 3/8-16 x 3" hex bolt. Add a 3/8" washer and insert into the hole as shown in Figure 5.36. Add 18, 3/8" washers to act as a spacer as shown. Add a 3/8" lock washer. Next, add a 3/8" hex nut and tighten.

You need about 3/4" of thread showing after the nut is tight. If its not remove or add washers as needed. Finally add a 3/8" washer.

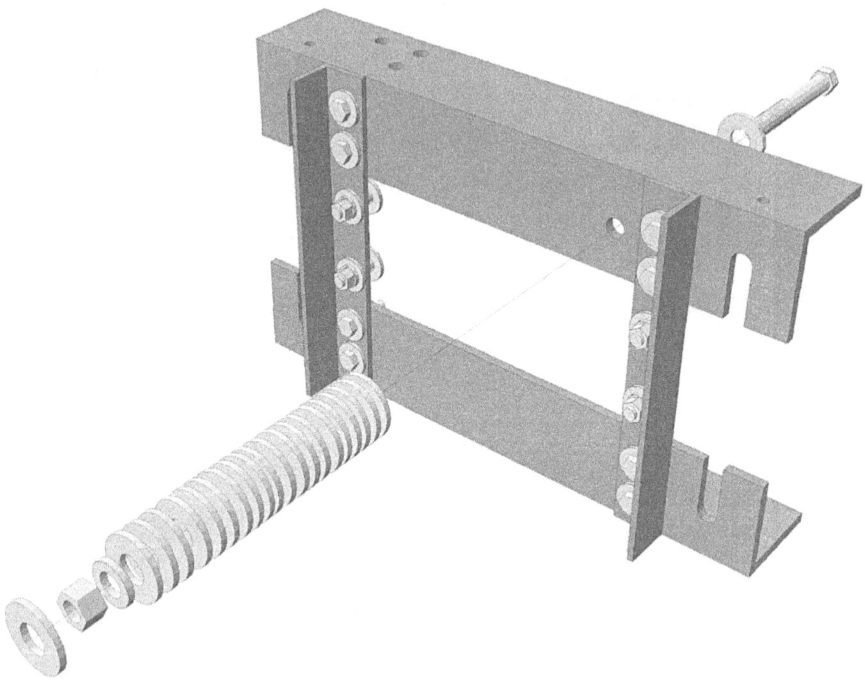

Figure 5.36

Step 18 - Attach the Y Dragon Cable

Lay the Y dragon cable with the head flush with the end of the side table, as shown in Figure 5.38.

Wrap the tail around and slip the 3/8" hole over the protruding 3/8" bolt as shown in Figure 5.37. Add a 3/8" washer, lock washer and hex nut. Position the tail as shown and tighten the nut.

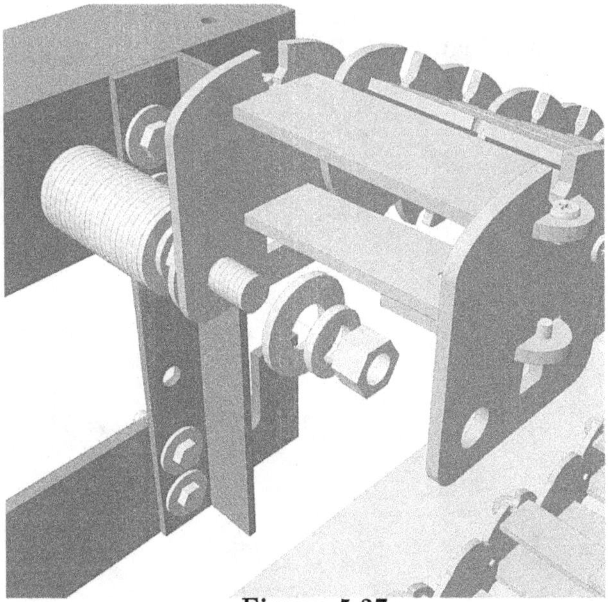

Figure 5.37

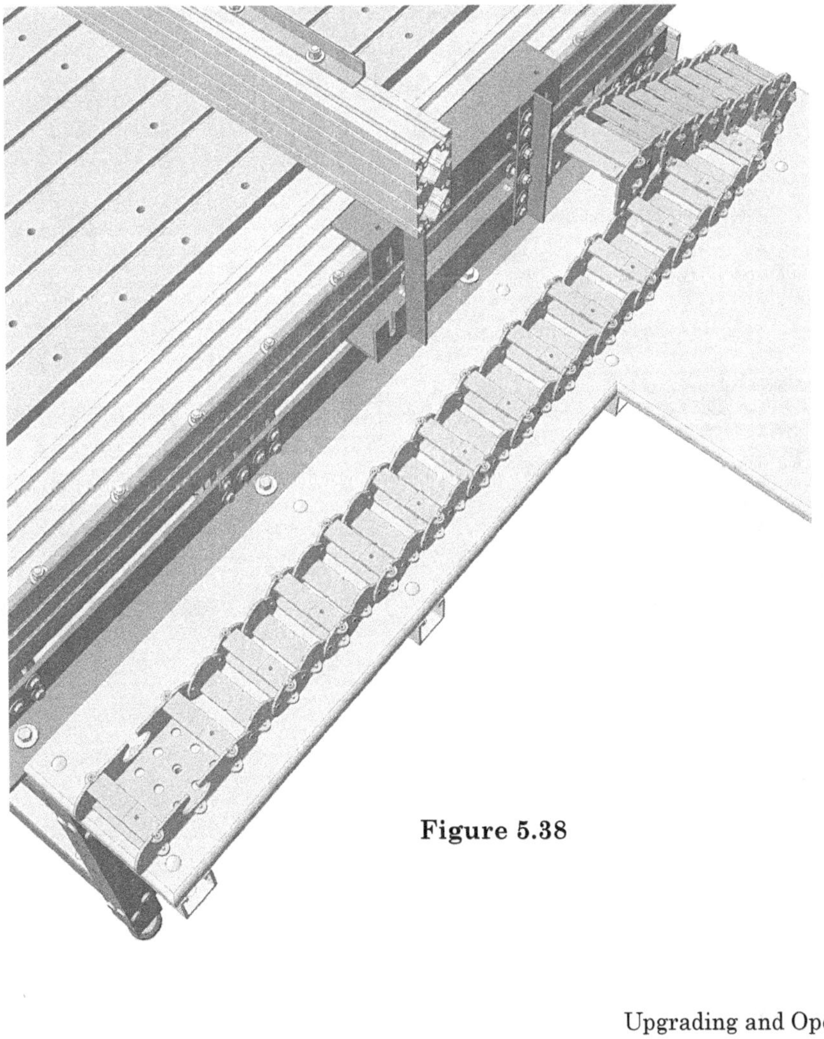

Figure 5.38

With the cable flush with the edge of the shelf, drill a 1/4" hole through the table in the position shown in Figure 5.39.

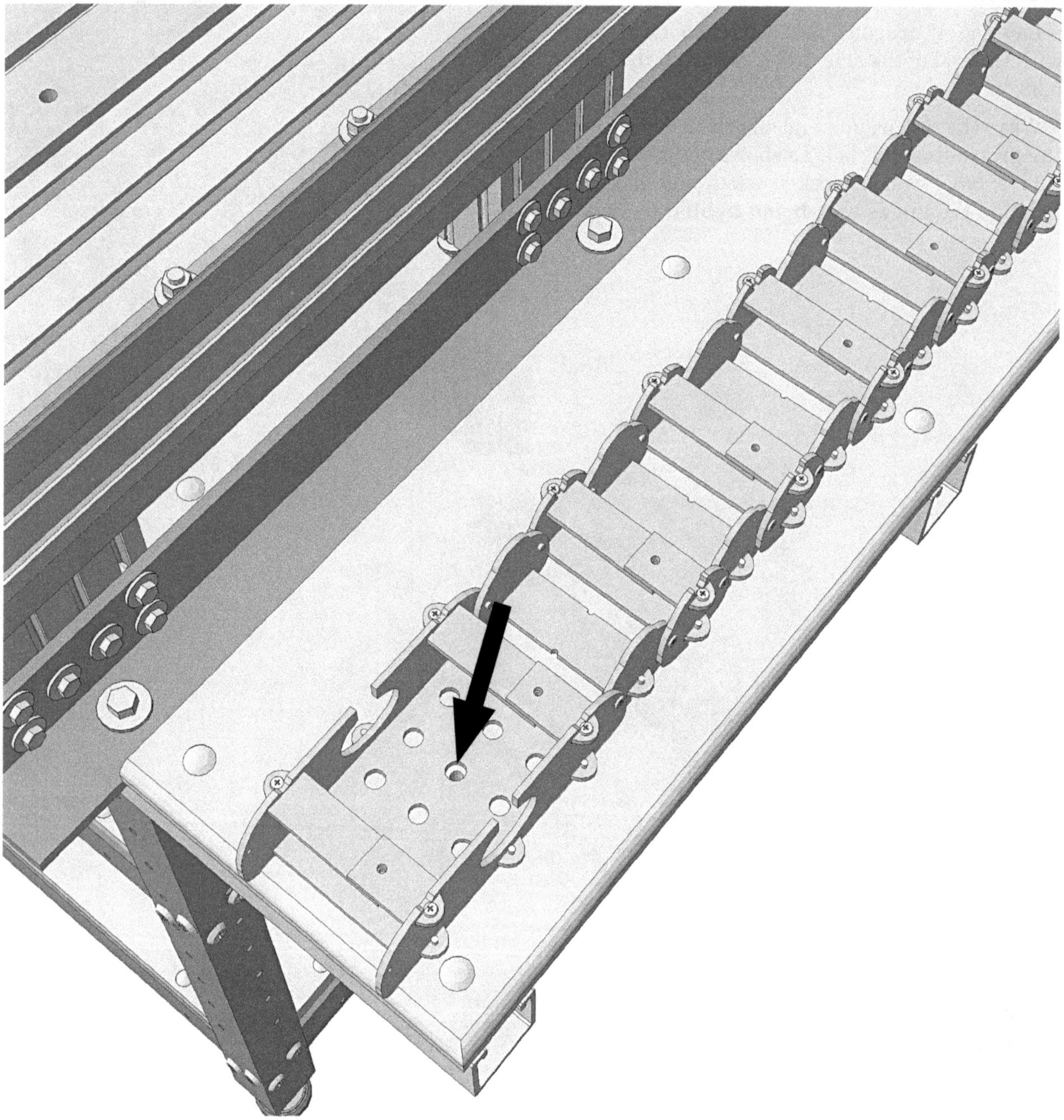

Figure 5.39

Move the cable out of the way and insert a 1/4-20 x 1-1/2" carriage bolt through the bottom of the side table, as shown in Figure 5.40. Add two 1/4" washers.

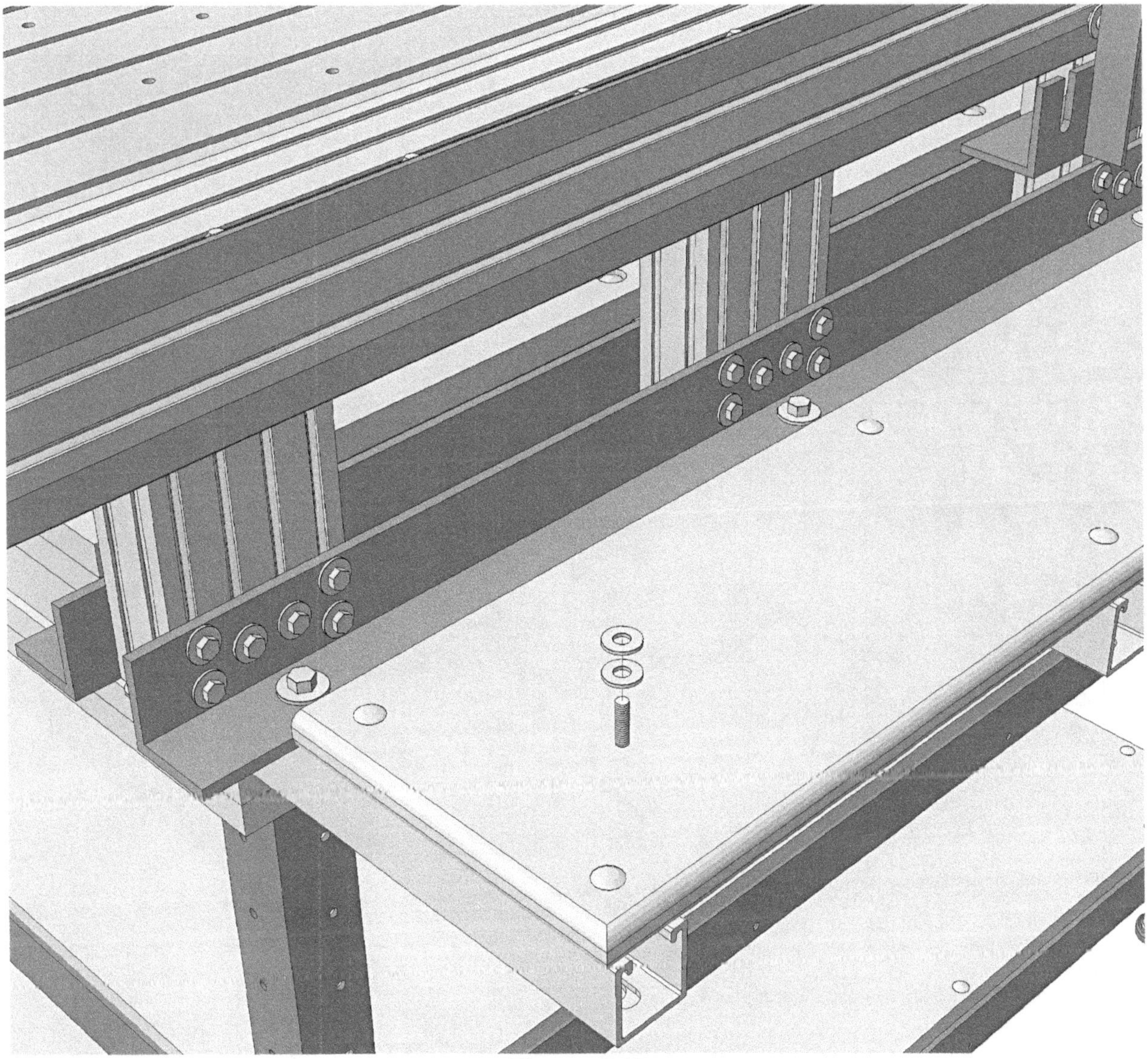

Figure 5.40

Slip the dragon cable over top of the carriage bolt and add a 1/4" washer, lock washer and hex nut, as shown in Figure 5.41. Tighten the nut.

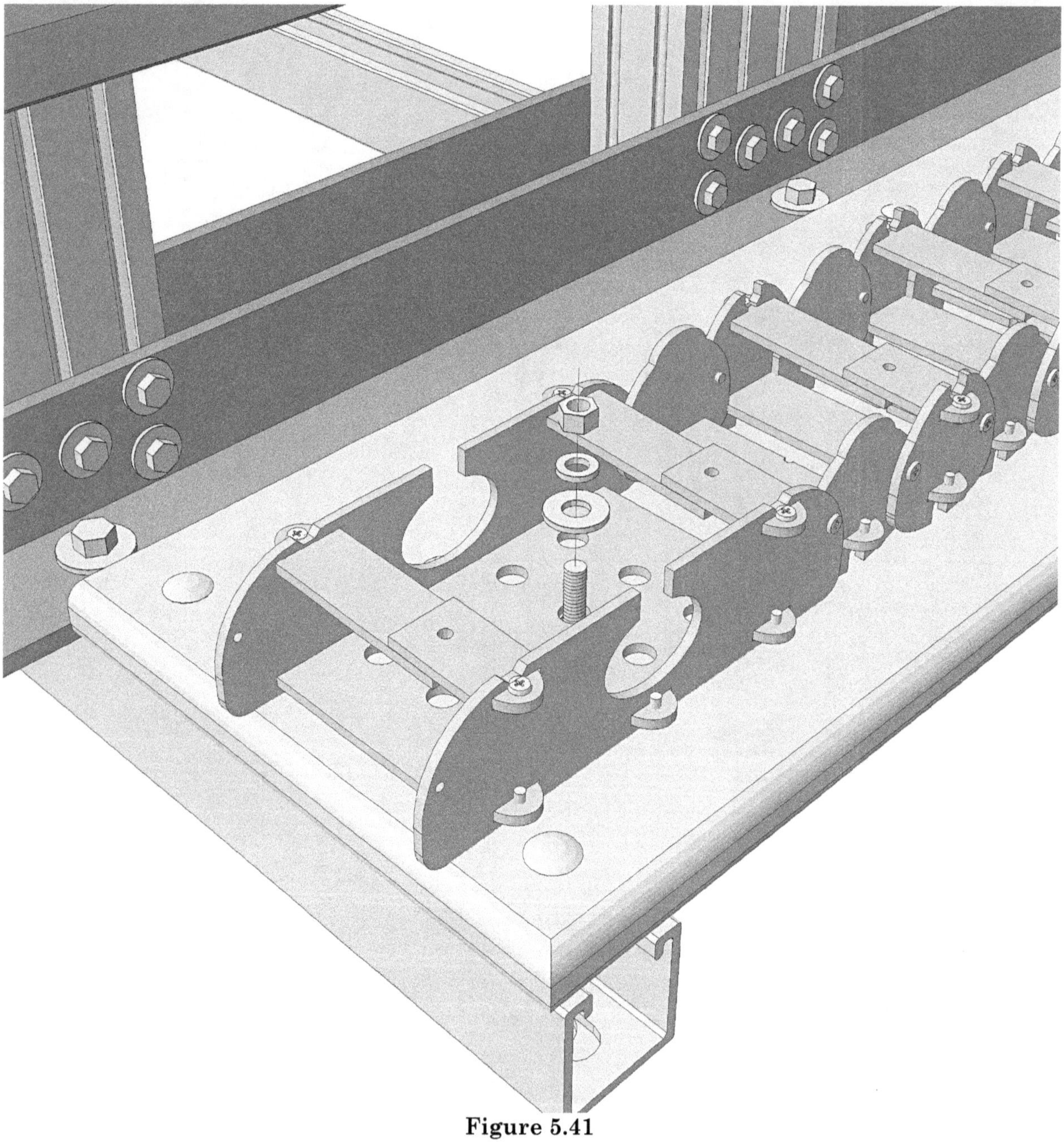

Figure 5.41

The finished dragon cables should look like those shown in Figure 5.42. All that's left is to run your cables.

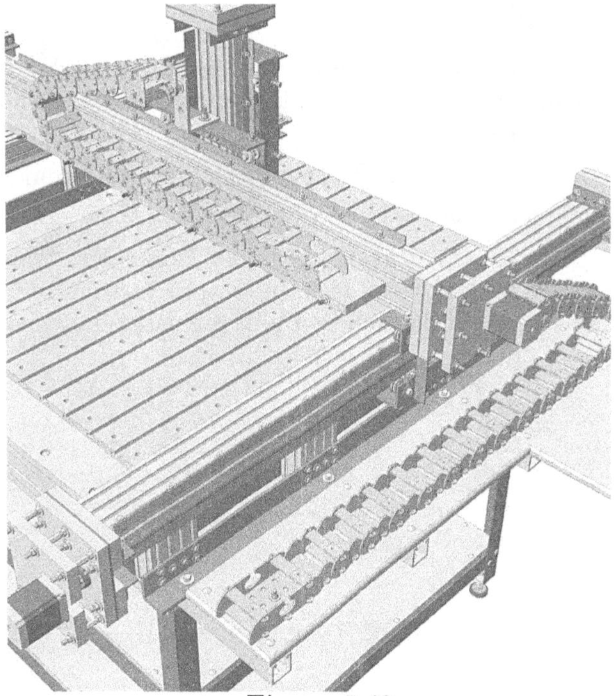
Figure 5.42

Step 19 - Run Cables

If your cables are currently attached to hanging supports remove them from the hangers

Rout them through the dragon cable. You insert a cable by carefully slipping it between the two upper supports as shown in Figure 5.43.

Figure 5.43

When routing cables start with the router power cable, routing it through the tail on the X dragon cable and around the bend, and finally exiting the head of the X cable. Then loop it around and into the tail of the Y dragon cable and finally out the head of the Y dragon cable.

Leave some wiggle room where the cable enters or exits a dragon cable. You want to give the cable enough room to enable securing the cable if necessary.

Once all the Z-carriage cables (router and Z axis motor) have been run move to the X axis motor cable.

Once the cables have been run slowly move the to the extent of all the axis while monitoring the cables.

Conclusion

Once all your cables have been run, add the 1/4" #6 machine screws to join the two upper inserts as shown in Figure 5.44.

It's best to hold the two pieces together with your fingers while aligning the holes when securing them.

Later you will be adding other upgrades. Some will require cables to be added. In most cases you can fish the cable through the links without removing the screws.

Figure 5.44

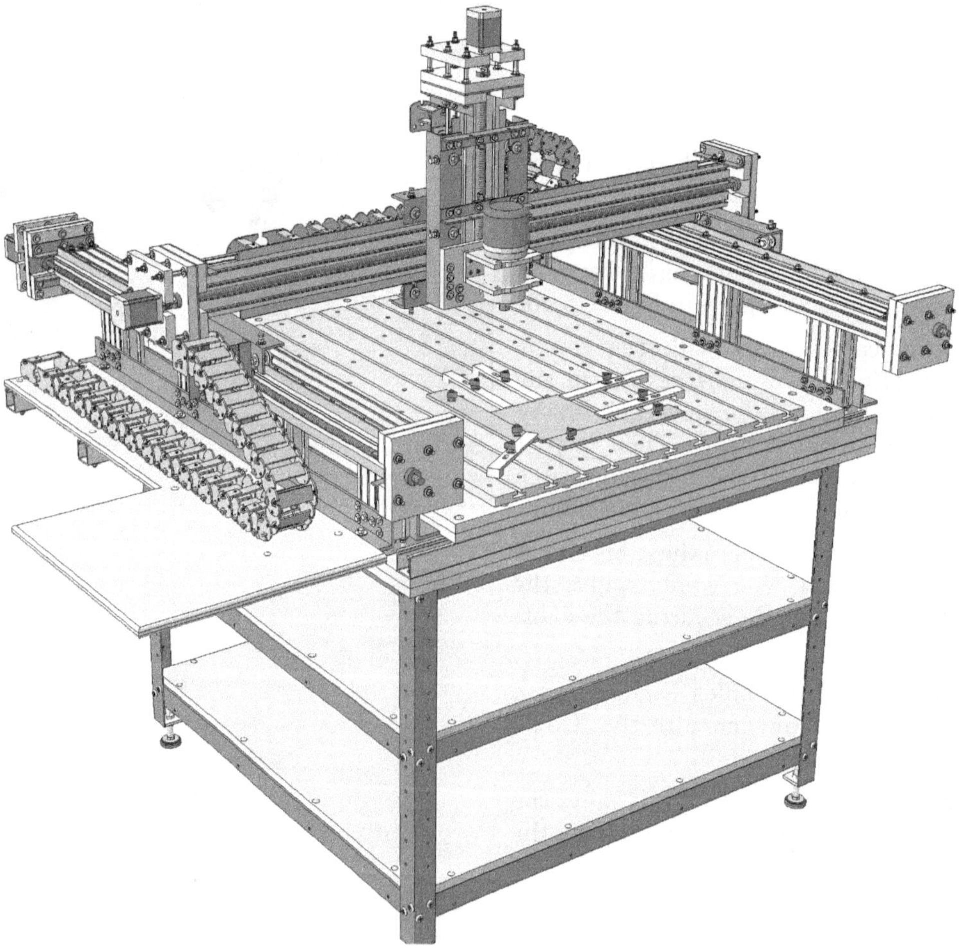

Chapter 6

Emergency Stop Switch

The emergency stop switch upgrade should be a high priority on your list of KRMx01 upgrades.

The emergency switch is wired directly into your G540 controller and when activated will cause the controller to remove power from the motors. It will also tell the Mach3 software to go into reset mode. There is no faster way to stop your CNC.

Tools Needed For This Chapter

- 7/16 Wrench or socket
- Drill press
- Freud 3/32” router bit
- Phillips screwdriver
- Sanding Sponge

Components Needed For This Chapter

- 1, 2’ x 4’ piece of tempered hardboard
- Emergency stop switch
- 10, #6 x 1/4 Machine screws
- 6, 1/4-20 Hex nuts
- 15’ of 2 conductor security cable.
- 1/2” Cable clamp

The cable can be just about any 16-24 gauge wire. Stranded will be easier to work with but solid will work if that’s all you can get. Cut about 6” off this cable. Later in the chapter you will use cutoff portion as hookup wire.

Time Estimates

Cost: $35

Time: One day

Prerequisites

You will need three pieces of 1/8" thick stock. 1/8" tempered hardboard is recommended. You will need the following sizes.

- Stop box front: 8" x 7"
- Stop box sides: 9" x 10"
- Stop box top: 8" x 5"

Don't forget you will need three pieces of waster board that is roughly the same size. Before starting the project cut all your stock and waster board to size.

The emergency stop switch is available at www.cncrouterparts.com. Its located in the Electronics section.

You will need a Freud 3/32" router bit shown in Figure 6.1. The part number for this bit is #04-098. You can purchase these from Amazon.

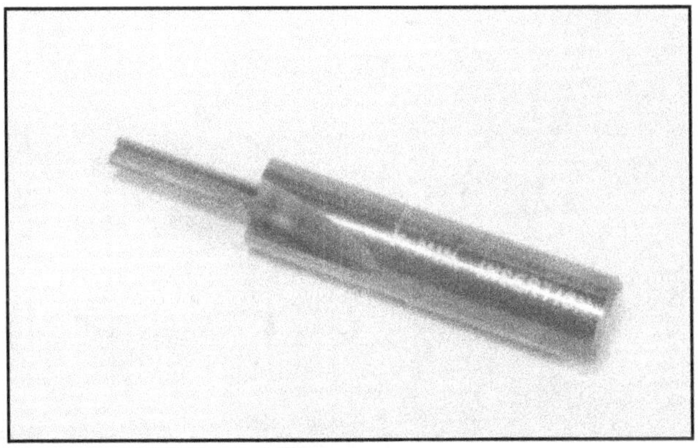

Figure 6.1

Important!

The power to your CNC should be removed during the installation of this upgrade.

Get the G-code files for this project here:

http://www.kronosrobotics.com/krmx01/da69781_9386_87112/

Step 1 - Cut the Stop Box Front

The first thing you need to do is cut your parts. Start with the front. Lay the 8" x 7" piece of stock and waster board on the CNC against your fences.

Important!

Make sure the 7 " side of the stock is along the Y axis (up and down) and the 8" side of the stock is along the X axis (left to right). Refer to the part orientation in Figure 6.2. Install the 3/32" router bit, zero the CNC and cut the part. Use the G-code file "Stop box front.txt"

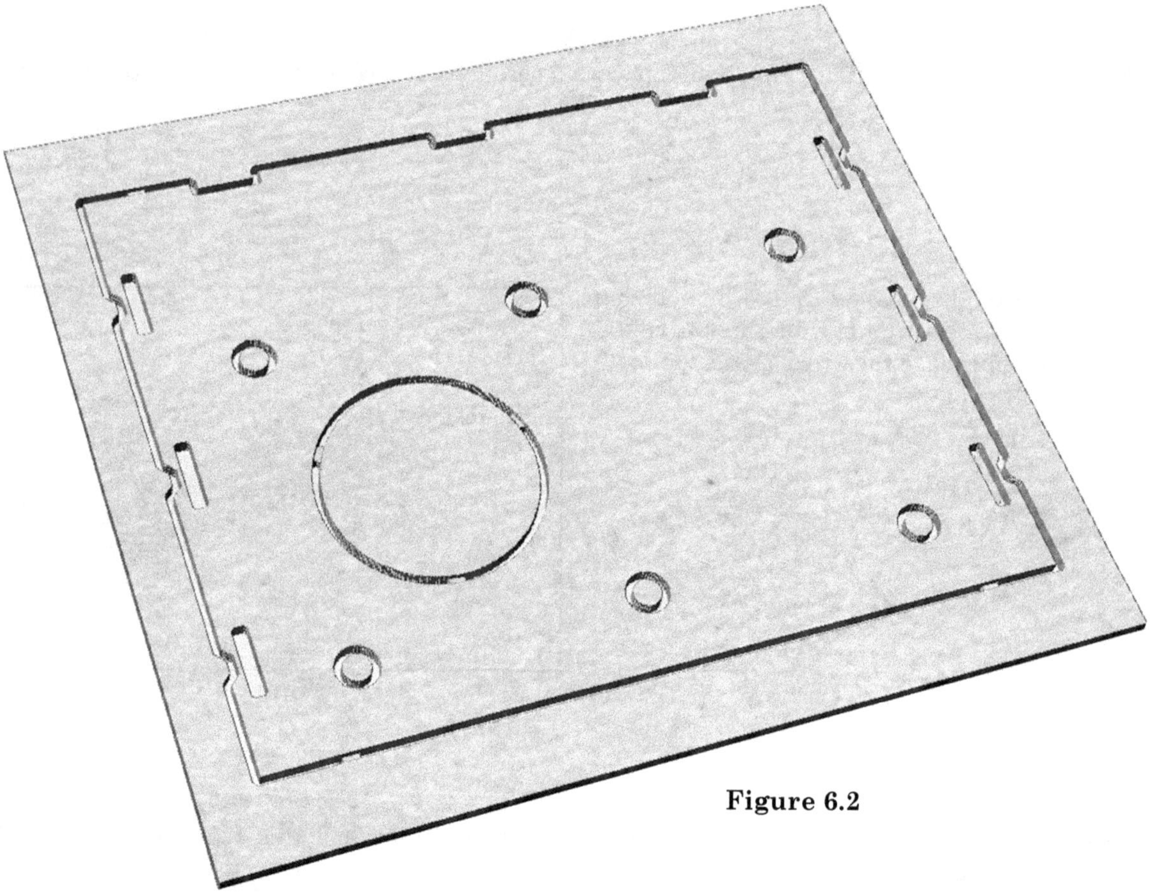

Figure 6.2

Step 2 - Cut the Stop Box Sides

Lay the 9" x 10" piece of stock and waster board on the CNC against your fences.

Important!

Make sure the 9" side of the stock is along the Y axis (up and down) and the 10" side of the stock is along the X axis (left to right). Double check this one as the long edge is up and down not side to side. Refer to the part orientation in Figure 6.3.

Make sure the CNC is zeroed and cut the part. Use the G-code file "Stop box side.txt"

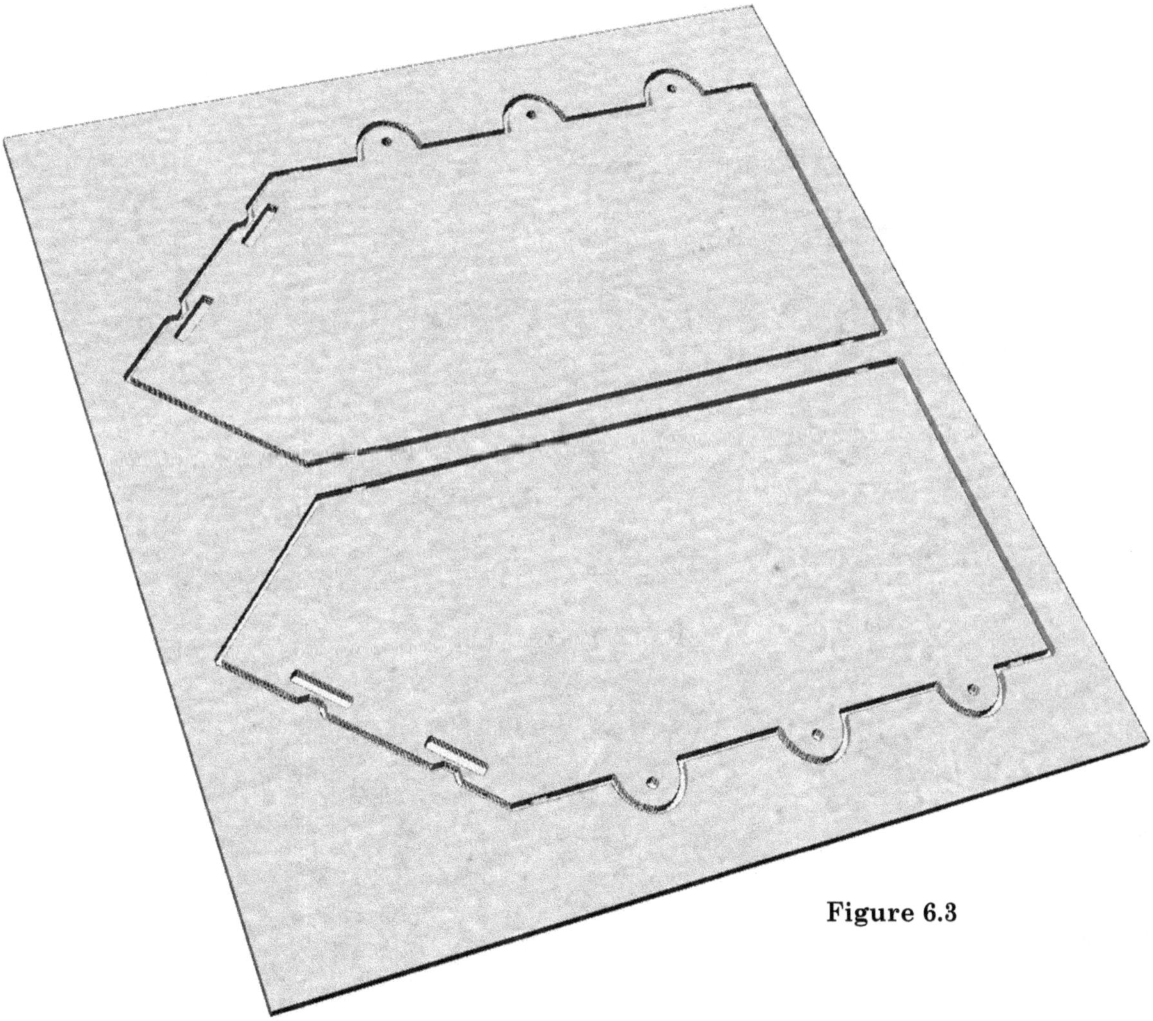

Figure 6.3

Step 3 - Cut the Stop Box Top

Lay the 8" x 5" piece of stock and waster board on the CNC against your fences.

Important!

Make sure the 5" side of the stock is along the Y axis (up and down) and the 8" side of the stock is along the X axis (left to right). Refer to the part orientation in Figure 6.4.

Make sure the CNC is zeroed and cut the part. Use the G-code file "Stop box top.txt"

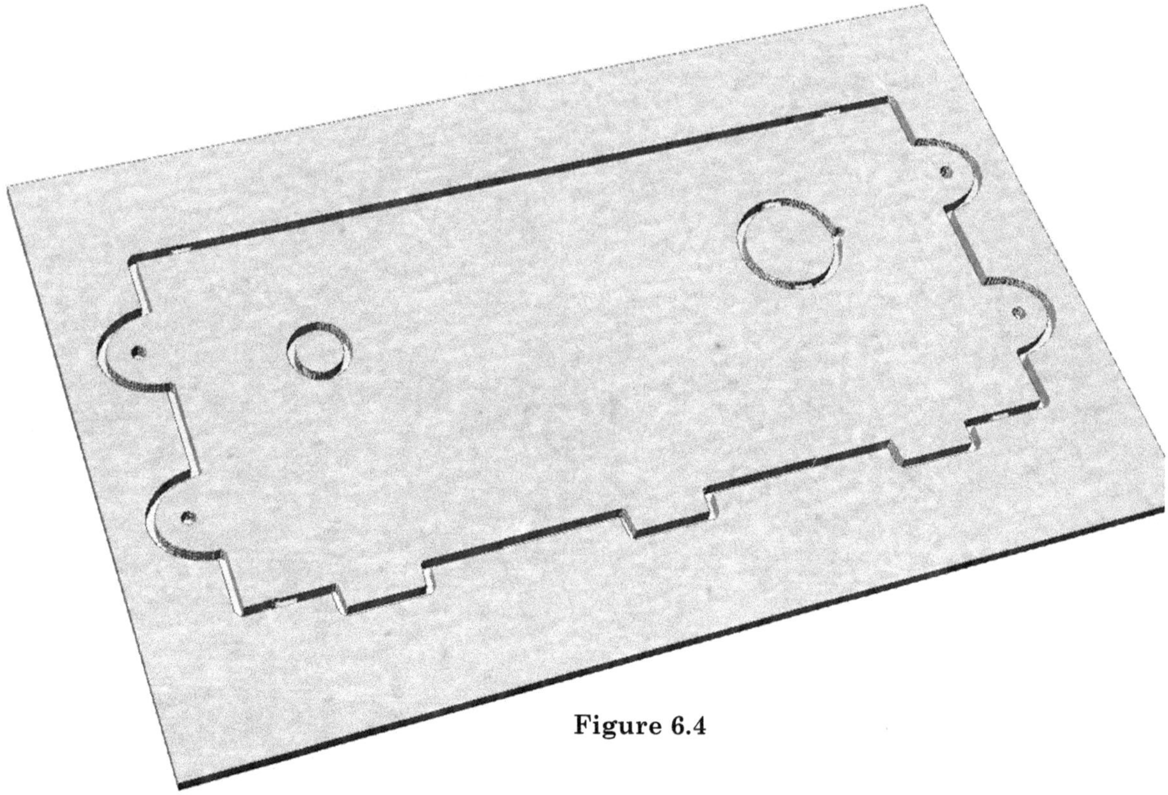

Figure 6.4

Step 4 - Clean the Parts

Once the parts have been cut, give them a light sanding with a sanding sponge while they are still attached to the stock

Step 5 - Remove Parts From Stock

When removing parts from the stock, always punch the main part out from the top. Punch internal cutouts from the underside.

Step 6 - Attach Sides to Top.

Insert the two sides onto the tabs of the top part as shown in Figure 6.5. Then using four, #6 x 1/4" machine screws, fix them in place. Don't overtighten the screws. Their job is to hold the sides pieces in place using a wedge action.

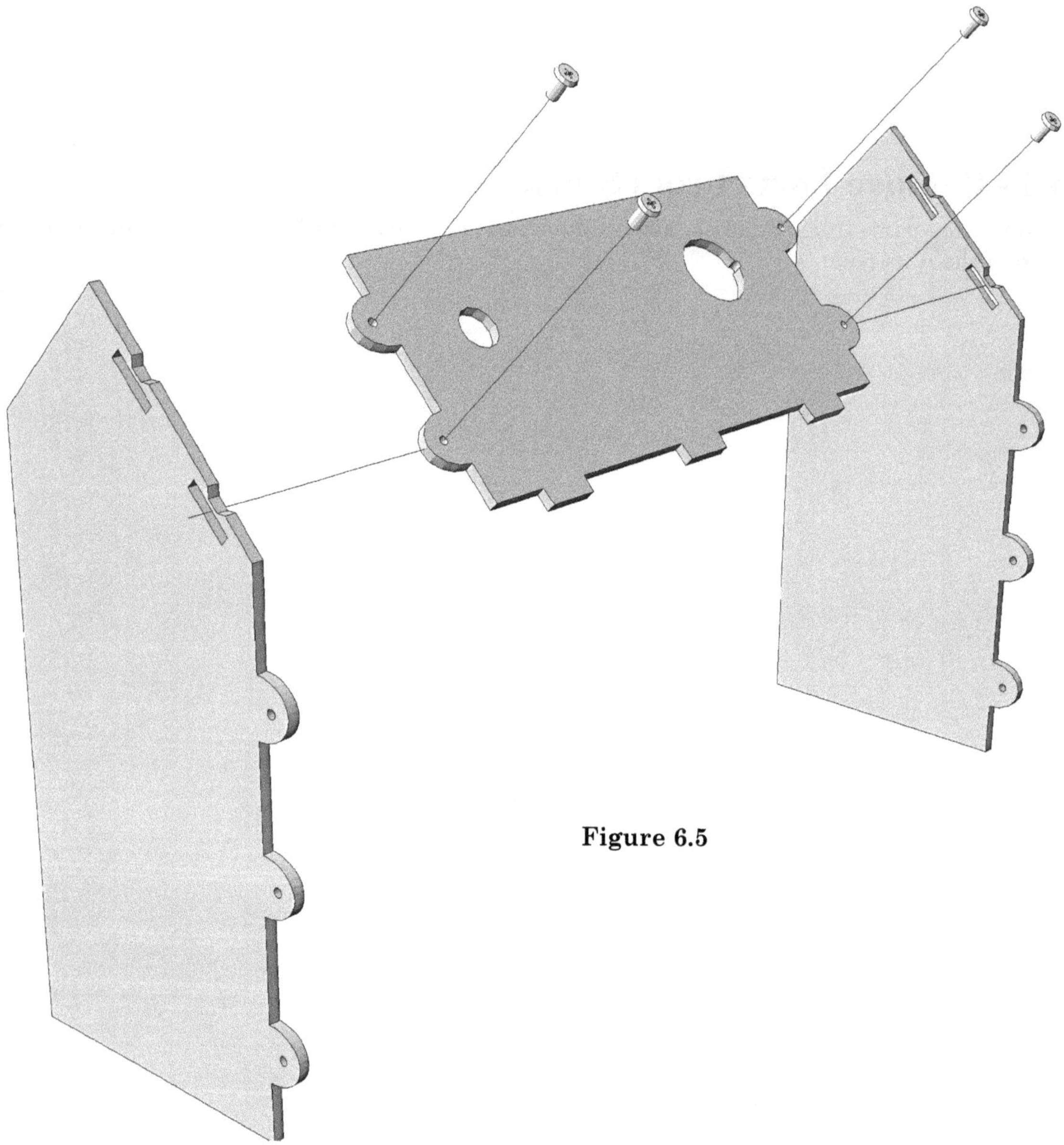

Figure 6.5

Step 7 - Attach Front

Insert the font part over the six tabs as shown in Figure 6.6. Fix in place with six #6 x 1/4" machine screws.

Once assembled, if you are going to paint your stop box you should do it now.

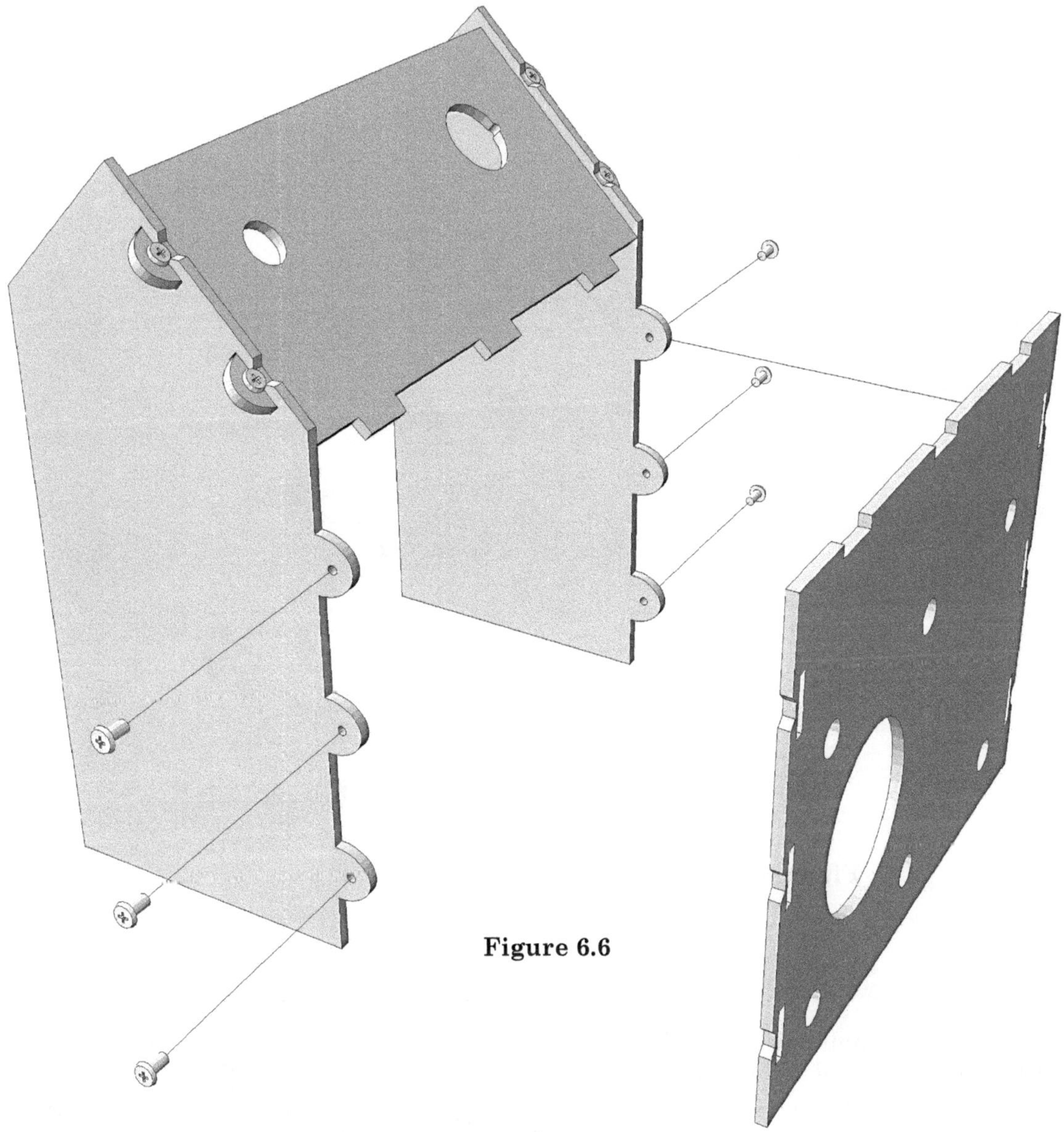

Figure 6.6

Step 8 - Install Switch Button

In order to install the switch, you need to remove the button portion from the switch housing. This is done by twisting the lever until it clicks. You twist it away from the writing on the housing. Once it clicks Pull the button portion out of the housing.

Remove the plastic nut and insert the button into the stop box as shown in Figure 6.7. Attach the plastic nut and tighten.

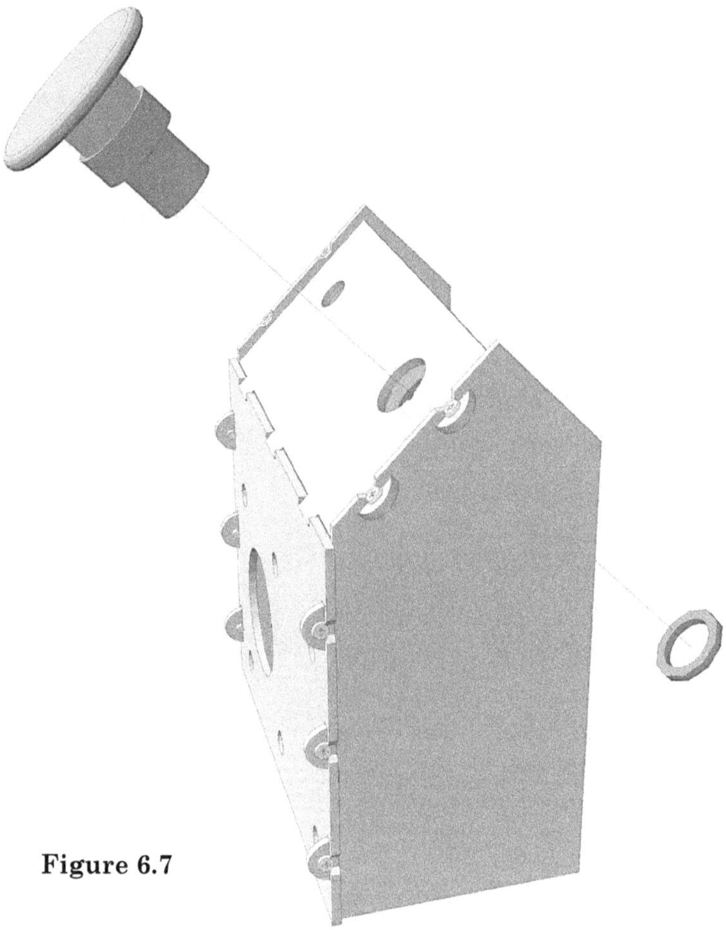

Figure 6.7

Tip

The button has a small notch located on one side near the top of the threads. This notch fits into the notch in the mounting hole on the stop box.

Step 9 - Attach Stop Box to CNC

Slip the stop box over the six bolts on the front left bearing block as shown in Figure 6.8. Hold in place with six 1/4-20 hex nuts.

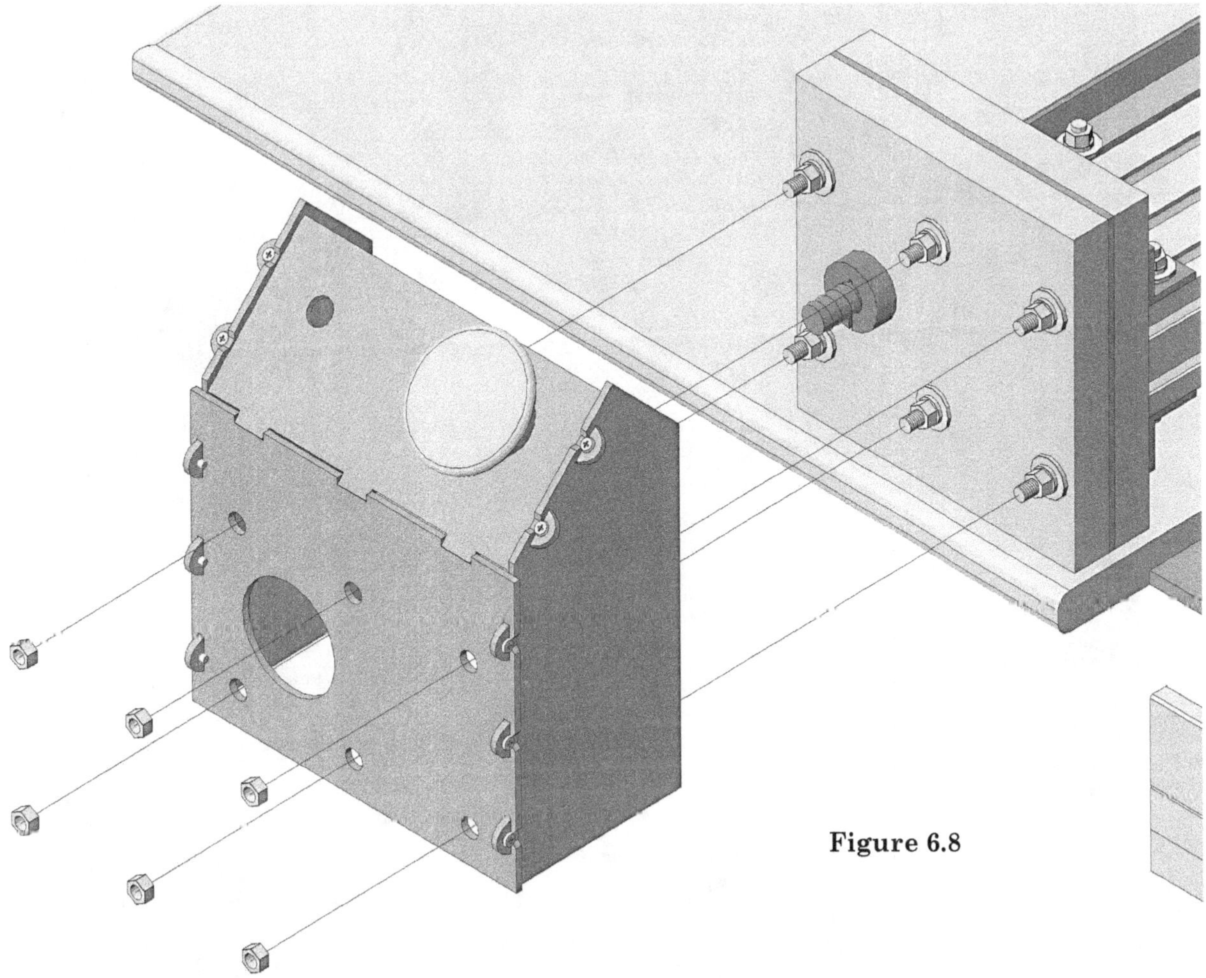

Figure 6.8

Step 10 - Attach Wires

Using a Phillips screwdriver, attach the two wires on one end of the 15' cable to the two screw terminals on the switch housing as shown in Figure 6.9. If you are using multi colored wire, you can ignore the colors.

Figure 6.9

Step 11 - Attach Switch Housing to Button

Attach the switch housing to the button by slipping the housing over the lower portion of the button located inside the EStop box, as shown in Figure 6.10. Snap it in place.

Figure 6.10

Step 12 - Run the Cable

Insert the cable into one of the slots in the Y-beam as shown in Figure 6.11. Use a small piece of craft foam to hold the cable in place. Note that the illustration shows four cables. These are for upgrades you will be adding to your CNC later. For now just use the center slot.

Figure 6.11

Tip

You small pieces of craft foam to hold the cable in place. You can use a color that matches that of your CNC.

When you installed the bearing blocks and motor mounts on the KRMx01, you were instructed to add some extra carriage bolts to the beam for future use. Add a cable clamp to one of those bolts to help hold the cable near the rear of the Y-beam, as shown in Figure 6.12. If you don't have a cable clamp, you can use two tie wraps.

Figure 6.12

Tie the cable to the cable that is attached the Y axis motor and rout it through the bottom shelf so that it routs over the top of your electronics board and out the front of the CNC.

Step 13 - Remove EStop Bypass wire.

In Book 1 of the KRMx01 series, you installed the EStop bypass wire shown in Figure 6.13. Remove this wire now.

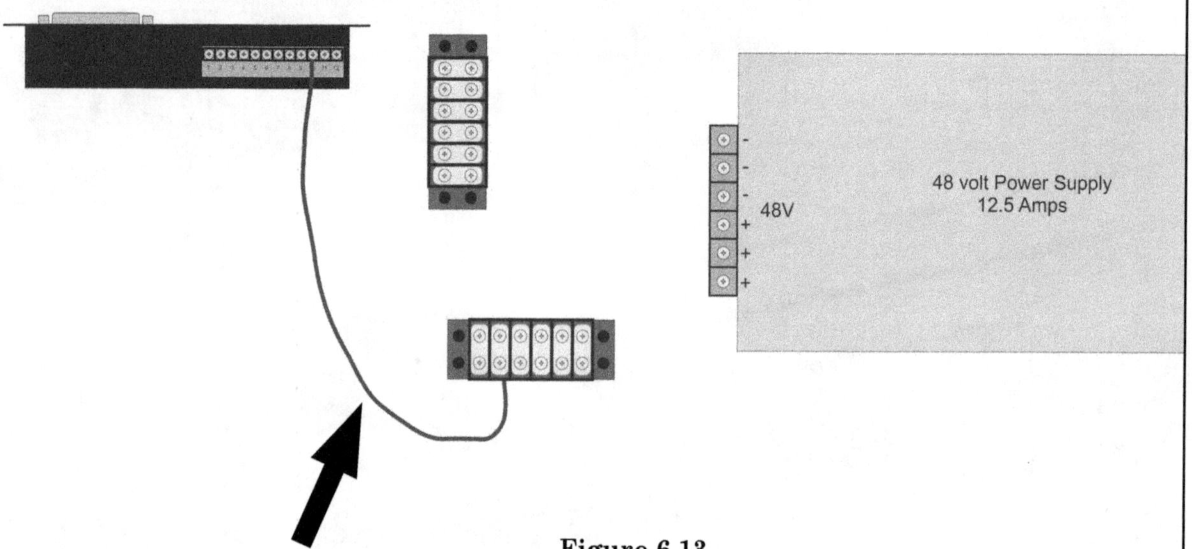

Figure 6.13

Step 14 - Attach Cable to Terminal Block

Strip about 1/2 of insulating from the wires on the end of the cable you just ran. Attach those wires to the terminal block shown in Figure 6.14. The color does not mater if you are using multi colored wire.

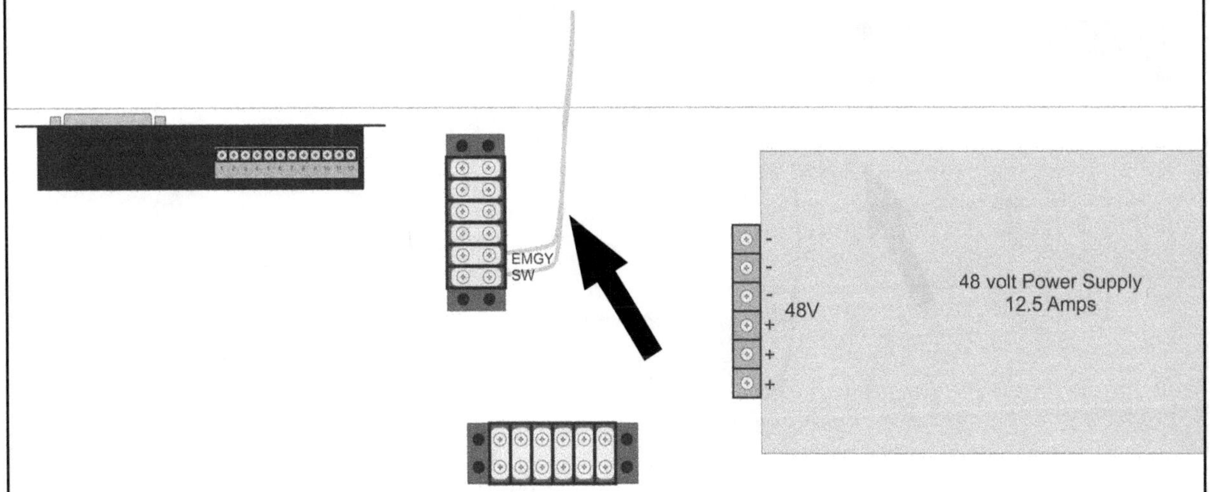

Figure 6.14

Step 15 - Attach to G540 Controller

Take a piece of hookup wire and connect it between terminal 10 on the G540 and the lower terminal on the terminal block, as shown in Figure 6.15. Connect another wire between the two terminal blocks as shown.

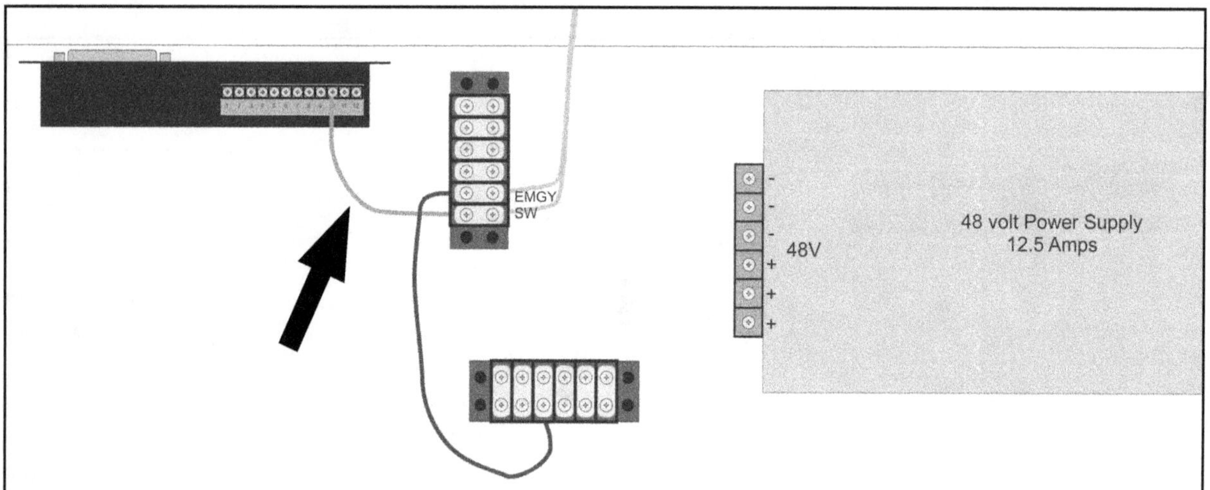

Figure 6.15

Tip

For instructional purposes, only wires that are pertinent to the instructions covered in this step are shown. Your actual wiring will contain other connections.

Step 16 - Check Your Wiring

Figure 1.16 shows all the wiring as it relates to this upgrade. Double check the wiring.

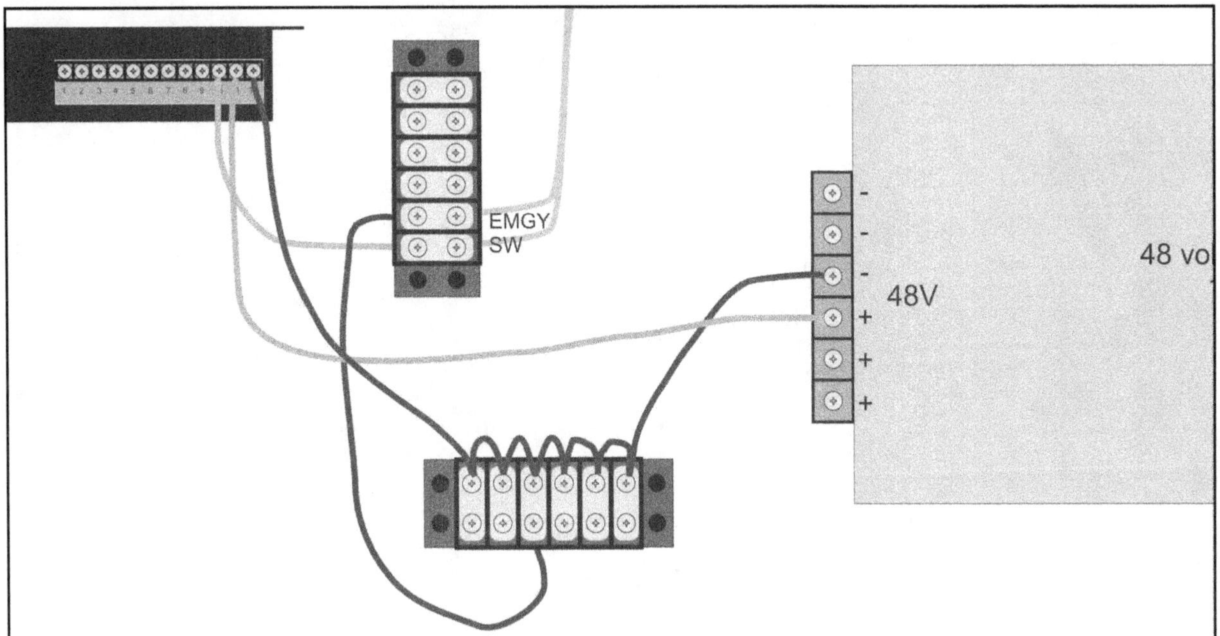

Figure 6.16

Operation

While hitting the Estop will shutdown the motors, you need to configure Mach3 so that it also knows that your EStop button has been hit. The G540 puts that information on Pin 15 of the DB25 connector. Edit the Ports and Pins Input Signals form so that the settings for the Estop are the same as shown in Figure 6.17.

The normal position of the emergency stop switch (EStop), is in the up position. When the button is pressed it will lock into the down position and cut the power to the motors. In order to return the button to normal operation mode (up), you need to twist the red portion of the button. When twisted it will pop back into the up position.

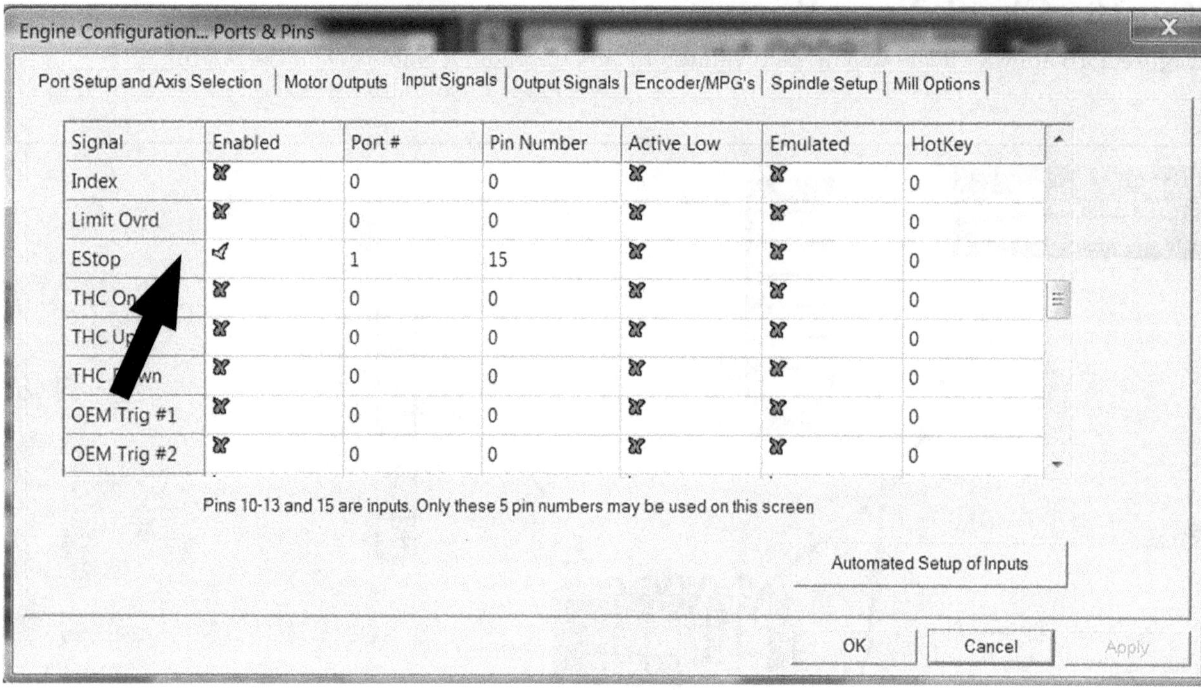

Figure 6.17

Switch Test

Make sure the switch is in the up position. Apply power to your CNC and boot your PC. Load Mach 3. You should hear the motors kick in once the program has started.

In Mach 3, hit the red reset button as you have done in the past. The button should turn green and your CNC should operate normally. If it does not you need to go back and check your wiring.

Now press the EStop button. The motors should disengage and Mach 3 should go into reset mode. To get back to normal operating mode, release the EStop button and reset Mach 3.

Tip

The EStop button will stop you CNC so fast that many times it will misstep and the steppers will loose sync with Mach 3. For this reason anytime you use your EStop switch you make sure you machine is still zeroed properly. Later when you install your homing switches you will re-home if you hit the EStop switch.-

Chapter 7

Homing Switches

A set of homing switches will allow you to keep accurate references to both the home position and extents of your machine. By adjusting your fences to your homed reference you can quickly and consistently setup your machine to cut various sizes of stock.

The homing switch upgrade is a massive upgrade. For this reason, the homing switch upgrade is broken down into four chapters. Each chapter will take you through the assembly and installation of each switch.

Each chapter will need one of the following switch assemblies. They can be purchased from the Kronos Robotics web site.

KRMx01 Y Homing Switch Assembly

http://www.kronosrobotics.com/xcart/product.php?productid=16590

KRMx01 A Homeing Switch Assembly

http://www.kronosrobotics.com/xcart/product.php?productid=16589

KRMx01 X Homing Switch Assembly

http://www.kronosrobotics.com/xcart/product.php?productid=16591

KRMx01 Z Homing Switch

http://www.kronosrobotics.com/xcart/product.php?productid=16592

In addition to the four switch chapters a chapter on wiring and configuration also included.

Chapter 8

X Homing Switch

This chapter will take you step-by-step through the assembly of the KRMx01 X homing switch kit assembly.

Tools Needed For This Chapter

- Phillips screwdriver
- Thread lock

Components Needed For This Assembly

- Kronos Robotics KRMx01 X homing switch kit. (see text)

The kit can be purchased from the Kronos Robotics web site here:
http://www.kronosrobotics.com/xcart/product.php?productid=16591

The kit consists of the following components.

- 23, Acrylic pieces
- 2, #6 x 1.5" Machine screws
- 2, #6 x 1" Machine screws
- 2, #6 x 3/4" Machine screws
- 3, #6 x 1/4" Machine screws
- 6, #6 Lock washers
- 6, #6 Hex nuts
- 2, #2 x 1/2" Machine screws
- 2, #2 Hex nuts
- 1, Micro switch with roller

Prerequisites

Later when you install the switch assemblies on your KRMx01 CNC, it is assumed that you added the extra 1/4-20 x 3/4" carriage bolts as outlined in the first KRMx01 book. These bolts will be used to secure your switch assembly. If you did not add these bolts, you will have to perform some disassembly to install the bolts.

Before starting the kit assembly be sure to remove the protective plastic from both sides of each of the acrylic pieces. Be sure all the holes are clear of debris that may still be present. The easiest way to do this is with the tip of a pencil or pen.

Warning

Acrylic is extremely brittle and until your switch assembly is assembled and installed there is a chance of breakage.

Tip

If you decide to create your own mounts for this homing switch, the wiring and configuration chapters can still be used to help you get them operational. You can purchase just the roller switch and screws by purchasing the Kronos Robotics Roller Switch kit.

http://www.kronosrobotics.com/xcart/product.php?productid=16612

Step 1 - Assemble Base

Sandwich the four acrylic components shown in Figure 8.1 together. Note that the component on the bottom has smaller threaded holes.

Insert the three #6 x 1/4" machine screws and tighten until they are snug only. They are only used to align the four pieces until they are attached to the CNC.

Place a small amount of thread lock on the tips of the three screws before inserting.

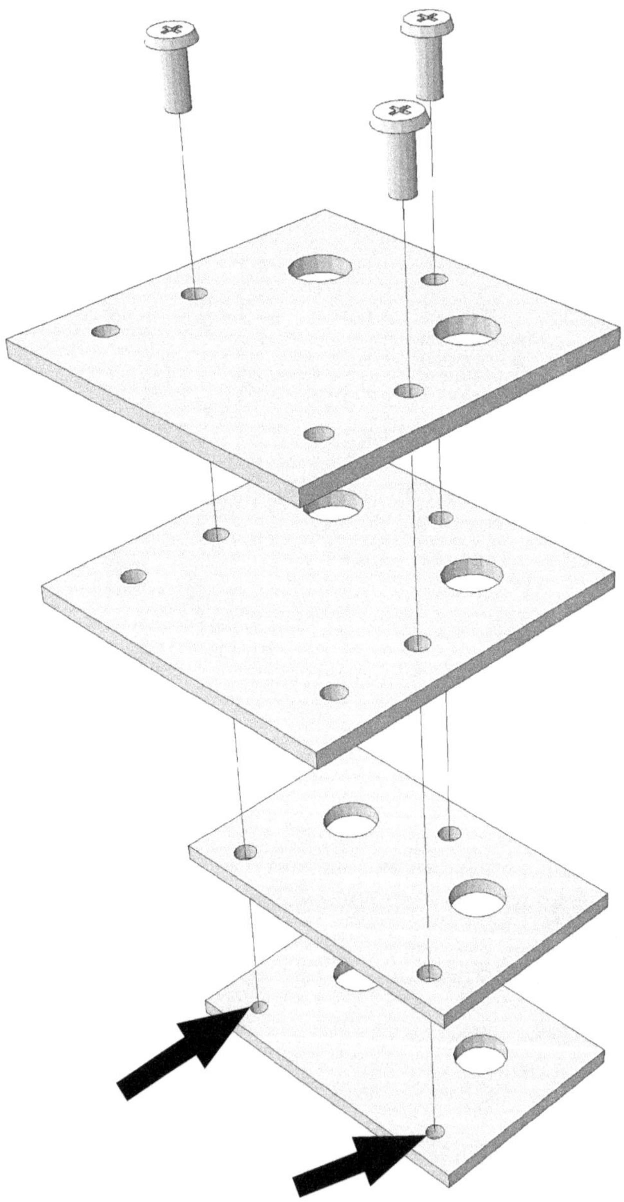

Figure 8.1

Step 2 - Attach Riser to Base

Take the three longer riser plates and insert the two #6 x 1-1/2" machine screws into the two center holes. Notice the orientation of the two outer holes, as shown in Figure 8.2.

Next, add ten of the smaller riser plates to the machine screws.

Insert the riser assembly into the base holes, again noting the orientation.

Add a #6 lock washer and nut to each machine screw and finger tighten.

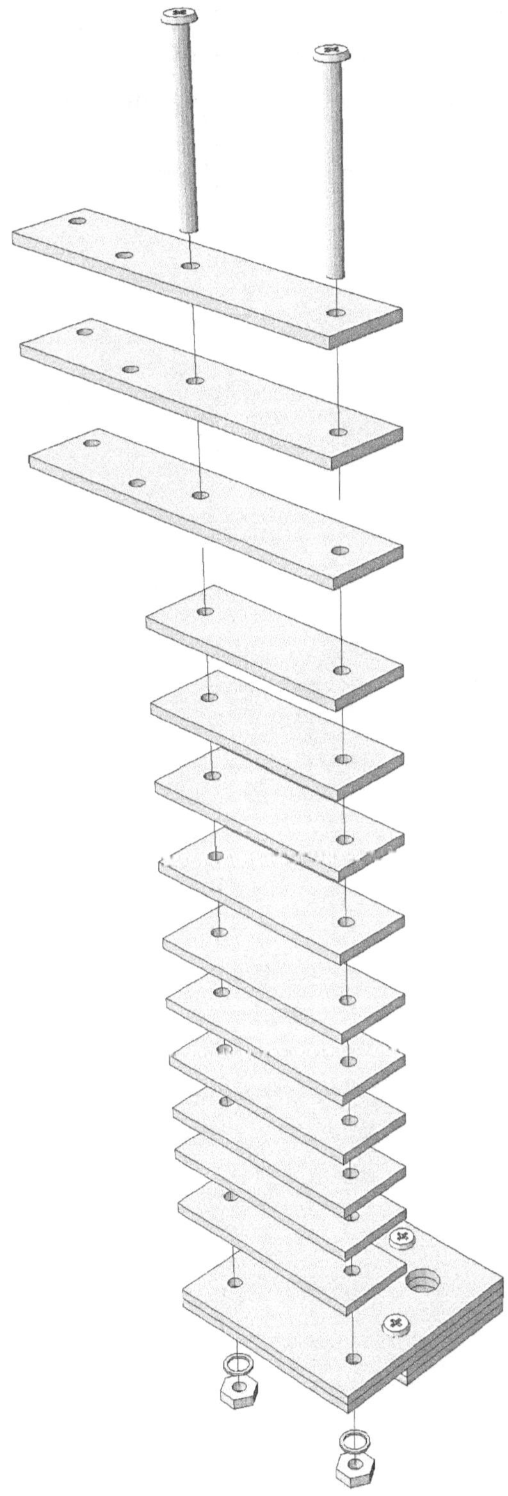

Figure 8.2

Step 3 - Assemble Top

Sandwich the six acrylic components shown in Figure 8.3 together. Notice the orientation of the holes.

Insert the two #6 x 3/4" machine screws into the two holes shown in Figure 8.3.

Add a #6 lock washer and hex nut and finger tighten.

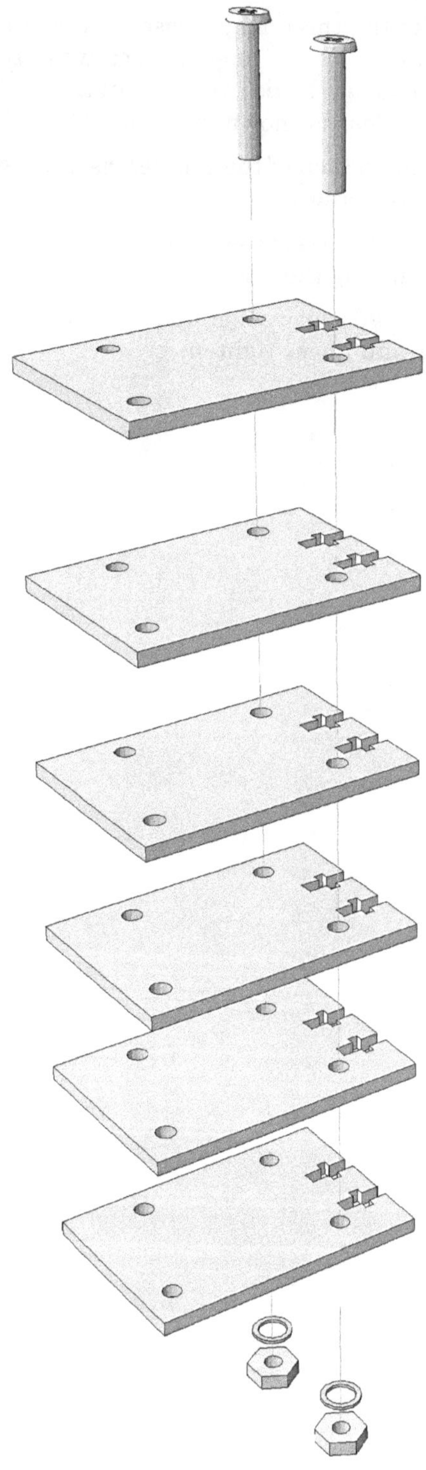

Figure 8.3

Step 4 - Attach Top to Assembly

Insert two #6 x 1" machine screws into the two holes shown in Figure 8.4. Then, insert the screws into the two matching holes on the long risers.

Add a #6 lock washer and hex nut and finger tighten.

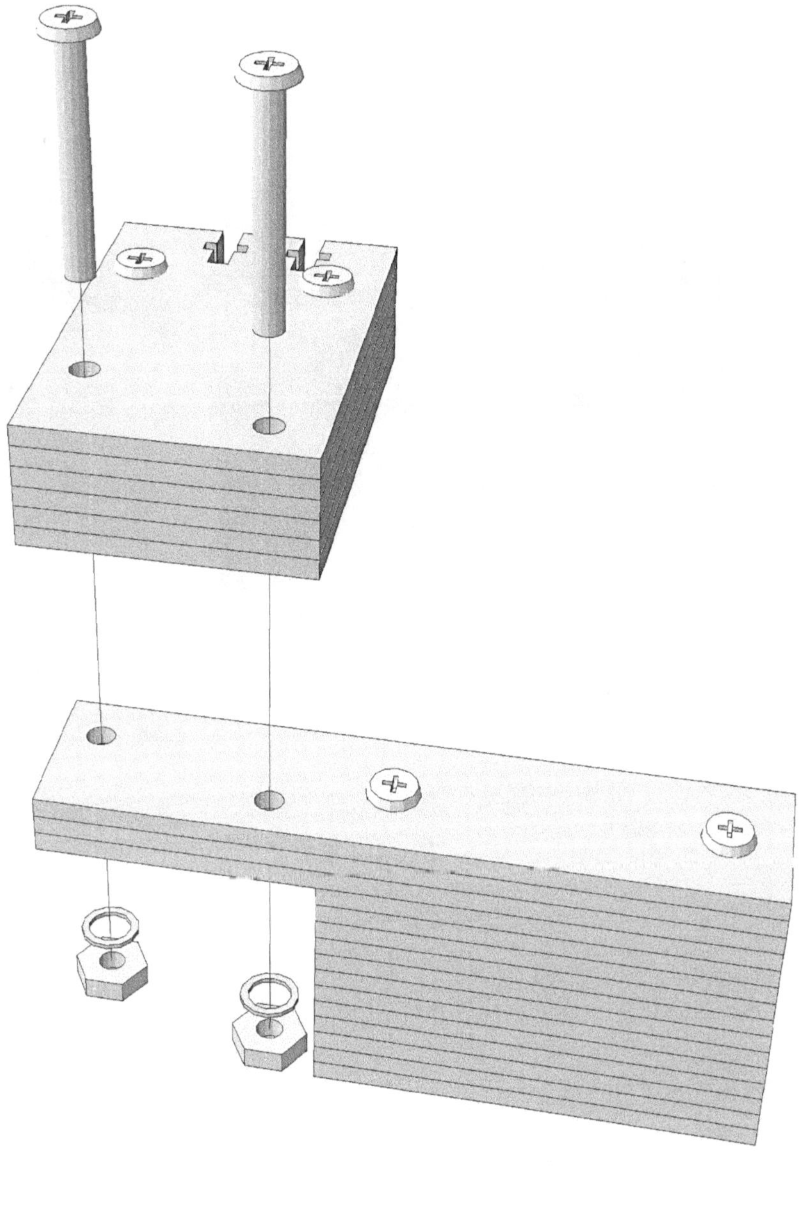

Figure 8.4

Step 5 - Assemble the Switch

Insert two #2 x 1/2" machine screws into the two holes on the micro switch, as shown in Figure 8.5. Pay attention to the orientation of the roller on the switch.

Add two #2 hex nuts to the ends of the screws. Do not tighten the nuts, leave them on the ends of the screws.

Add a couple drops of thread lock to the nuts.

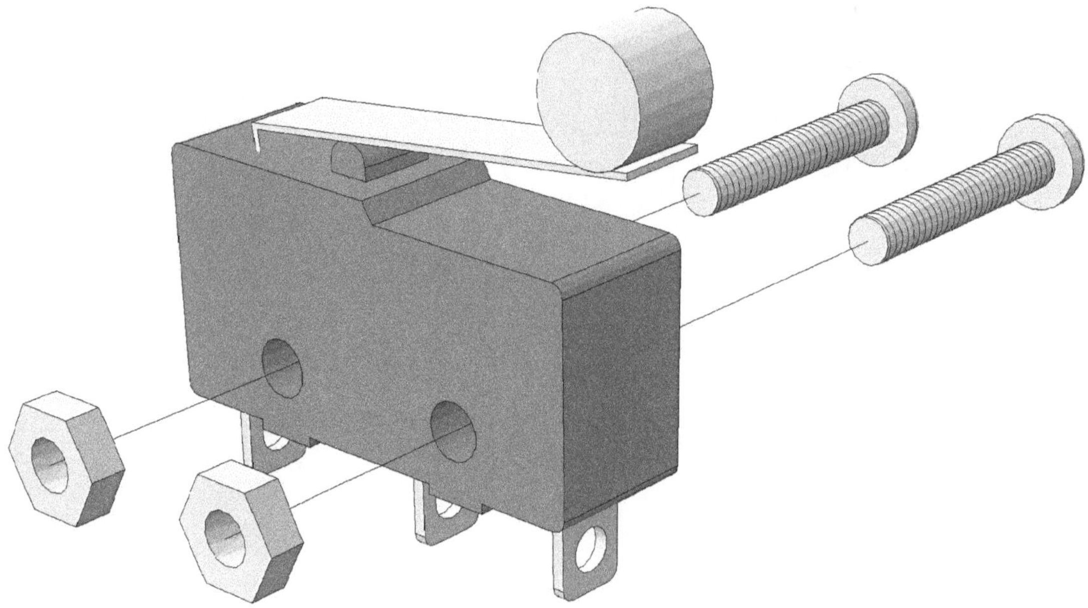

Figure 8.5

Step 6 - Attach the Switch

Slip the ends of the screws and nuts on the switch assembly into the slots of the top plates, as shown in Figure 8.6. Holding the plastic portion of the switch flush with the top plates. Tighten the two machine screws until the switch is held firmly in place.

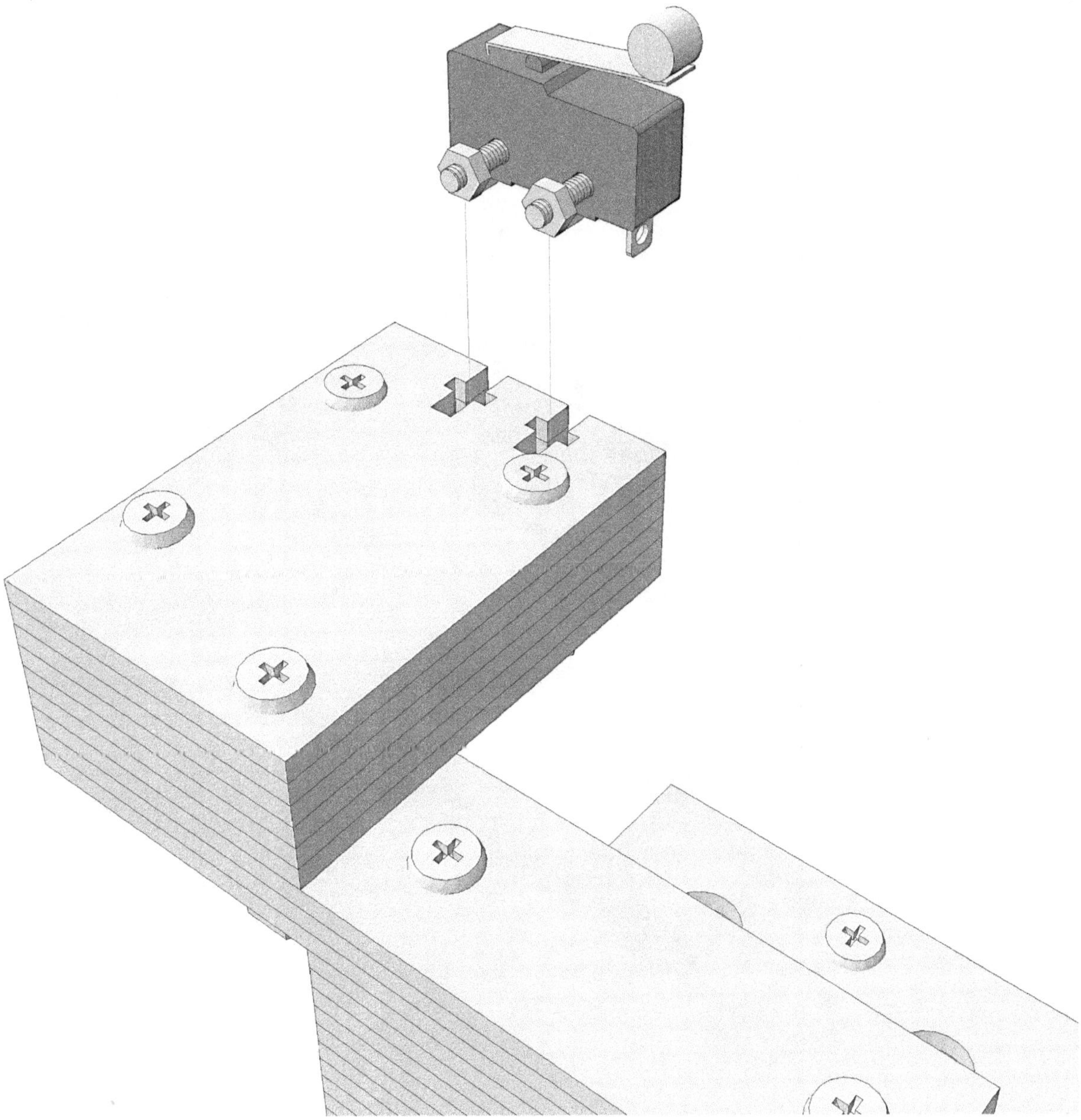

Figure 8.6

The final assembly should look like the one shown in Figure 8.7. Tighten the six machine screws that are attached to nuts. Don't over tighten or you may crack the acrylic. Tighten them enough so that the plates are held together firmly and do not shift.

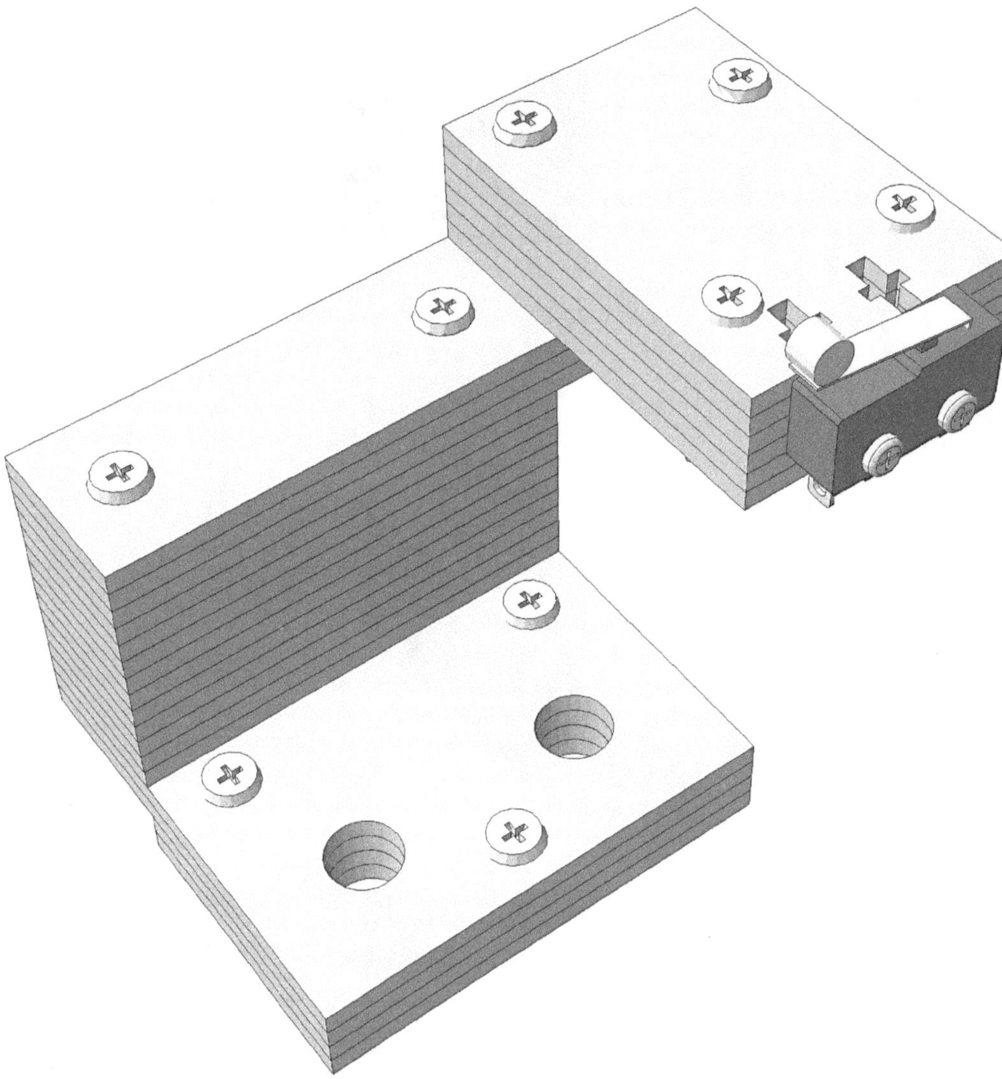

Figure 8.7

X Axis Homing Switch Installation

Step 1 - Attach Cable

You will need about 15' of cable to attach the X axis switch to your G540 controller. Strip 1/4" of the insulation from the two conductors on one end of the cable.

Optional: If you like, you can slip two small pieces of heat shrink tubing over the wires before soldering in place. 1/8" tubing works well.

Slip the bare ends of the wire through the two holes in the switch contacts. Notice the switch lead orientation. You want to connect the two wires to the Normally Opened contacts. These are marked NO and C.

Don't worry about the color of the wire connected as polarity does not matter. Solder the wires in place, as shown in Figure 8.8.

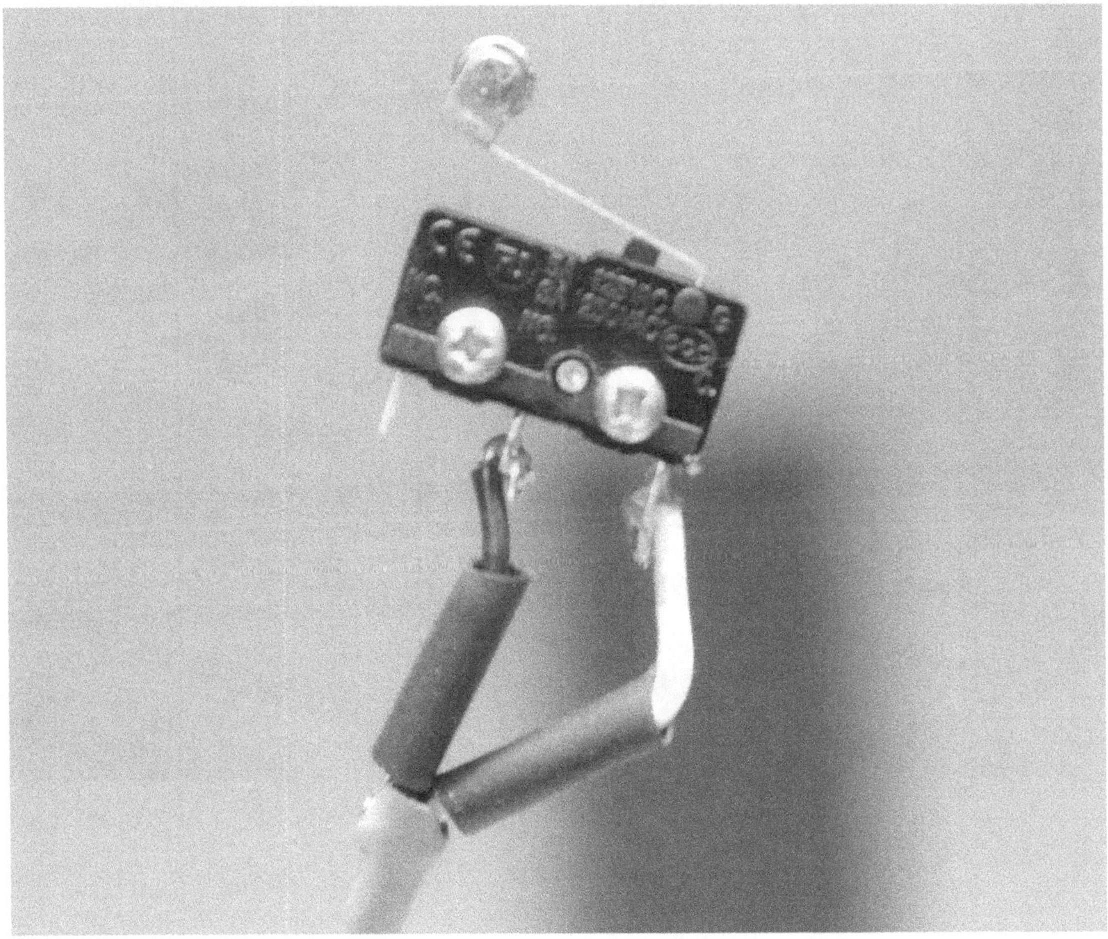

Figure 8.8

Step 2 - Optional

If you did install heat shrink tubing, slide it down over the contacts and apply a small amount of heat until the tubing shrinks around the connectors.

Step 3 - Attach Switch Mount

First place a 1/4" washer on each of the accessory bolts , then slip the switch mount assembly over the two bolts. Place a drop or two of thread lock on the end of each bolt.

Next, add a 1/4" washer and hex nut to each bolt, as shown in Figure 8.9.

The bottom of the switch mount should be 3-3/4" from the edge of the extrusion. Finger tighten the nuts.

Figure 8.9

Step 4 - Attach Switch to Mount

If you have not done so already, attach the switch to the mount as shown in Figure 8.10.

Adjust the switch up and down so that the X carriage can travel over the roller, slightly depressing it.

Figure 8.10

Conclusion

You want the X switch to kick in (you can here a click) with the X carriage as close to the Y-beam as possible without touching.

Figure 8.11

Chapter 9

Y Homing Switch

This chapter will take you step-by-step through the assembly of the KRMx01 Y homing switch kit assembly.

Tools Needed For This Chapter

- Phillips screwdriver
- Thread lock

Components Needed For This Assembly

- Kronos Robotics KRMx01 Y homing switch kit. (see text)

The kit can be purchased from the Kronos Robotics web site here:
http://www.kronosrobotics.com/xcart/product.php?productid=16590

The kit consists of the following components.

- 23, Acrylic pieces
- 2, #6 x 1.5" Machine screws
- 2, #6 x 1" Machine screws
- 3, #6 x 1/4" Machine screws
- 4, #6 Lock washers
- 4 #6 Hex nuts
- 2, #2 x 1/2" Machine screws
- 2, #2 Hex nuts
- 1, Micro switch with roller

Prerequisites

Later when you install the switch assemblies on your KRMx01 CNC, it is assumed that you added the extra 1/4-20 x 3/4" carriage bolts as outlined in the first KRMx01 book. These bolts will be used to secure your switch assembly. If you did not add these bolts, you will have to perform some disassembly to install the bolts.

Before starting the kit assembly be sure to remove the protective plastic from both sides of each of the acrylic pieces. Be sure all the holes are clear of debris that may still be present. The easiest way to do this is with the tip of a pencil or pen.

Warning

Acrylic is extremely brittle and until your switch assembly is assembled and installed there is a chance of breakage.

Tip

If you decide to create your own mounts for this homing switch, the wiring and configuration chapters can still be used to help you get them operational. You can purchase just the roller switch and screws by purchasing the Kronos Robotics Roller Switch kit.

http://www.kronosrobotics.com/xcart/product.php?productid=16612

Step 1 - Assemble Base

Sandwich the four acrylic components shown in Figure 9.1 together. Note that the component on the bottom has smaller threaded holes.

Insert the three #6 x 1/4" machine screws and tighten until they are snug only. They are only used to align the four pieces until they are attached to the CNC.

Place a small amount of thread lock on the tips of the three screws before inserting.

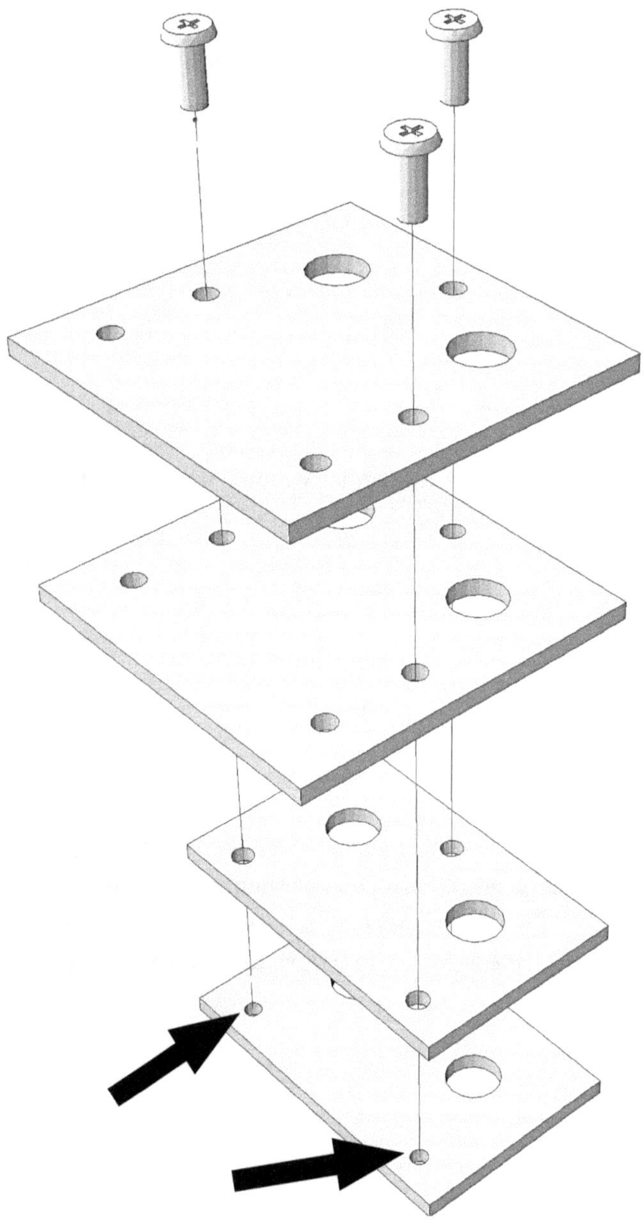

Figure 9.1

Step 2 - Attach Riser to Base

Take the three longer riser plates and insert the two #6 x 1-1/2" machine screws into the two center holes. Notice the orientation of the two outer holes, as shown in Figure 9.2.

Next, add ten of the smaller riser plates to the machine screws.

Insert the riser assembly into the base holes, again noting the orientation.

Add a #6 lock washer and nut to each machine screw and finger tighten.

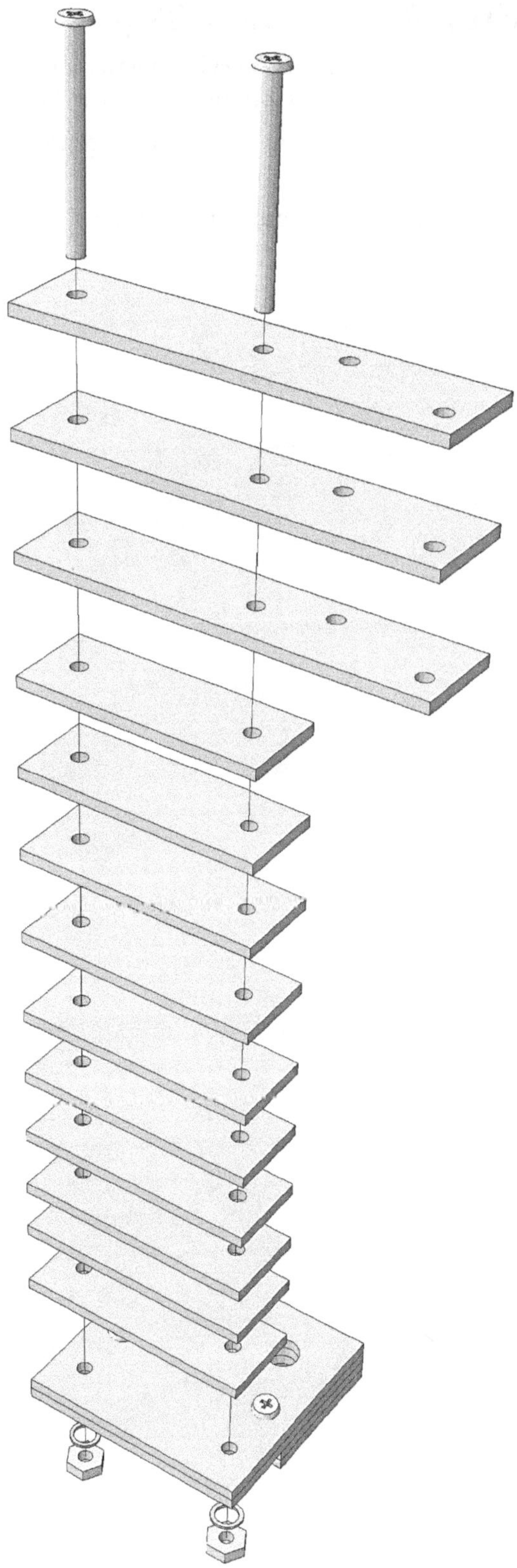

Figure 9.2

Step 3 - Assemble Top

Sandwich the six acrylic components shown in Figure 9.3 together. Notice the orientation of the holes.

Insert the two #6 x 1" machine screws into the two holes shown in Figure 9.3.

Don't attach any nuts at this time.

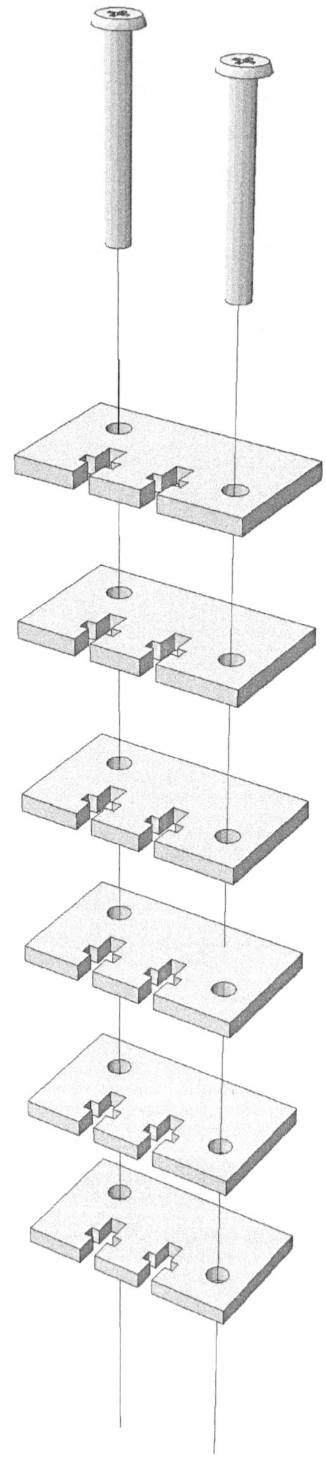

Figure 9.3

Step 4 - Attach Top to Assembly

Insert the two #6 x 1" machine screws into the two matching holes on the long risers.

Add a #6 lock washer and hex nut and finger tighten.

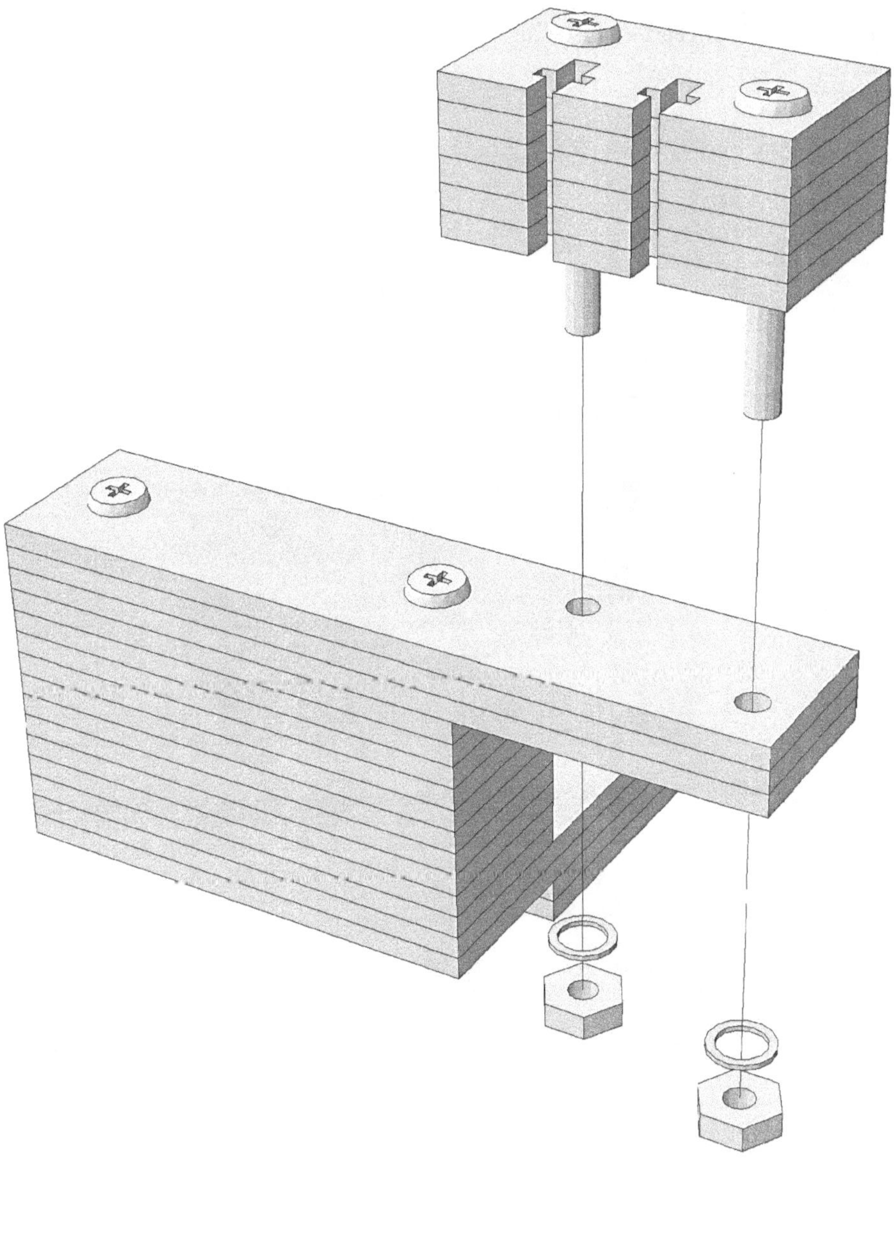

Figure 9.4

Step 5 - Assemble the Switch

Insert two #2 x 1/2" machine screws into the two holes on the micro switch, as shown in Figure 9.5. Pay attention to the orientation of the roller on the switch.

Add two #2 hex nuts to the ends of the screws. Do not tighten the nuts, leave them on the ends of the screws.

Add a couple drops of thread lock to the nuts.

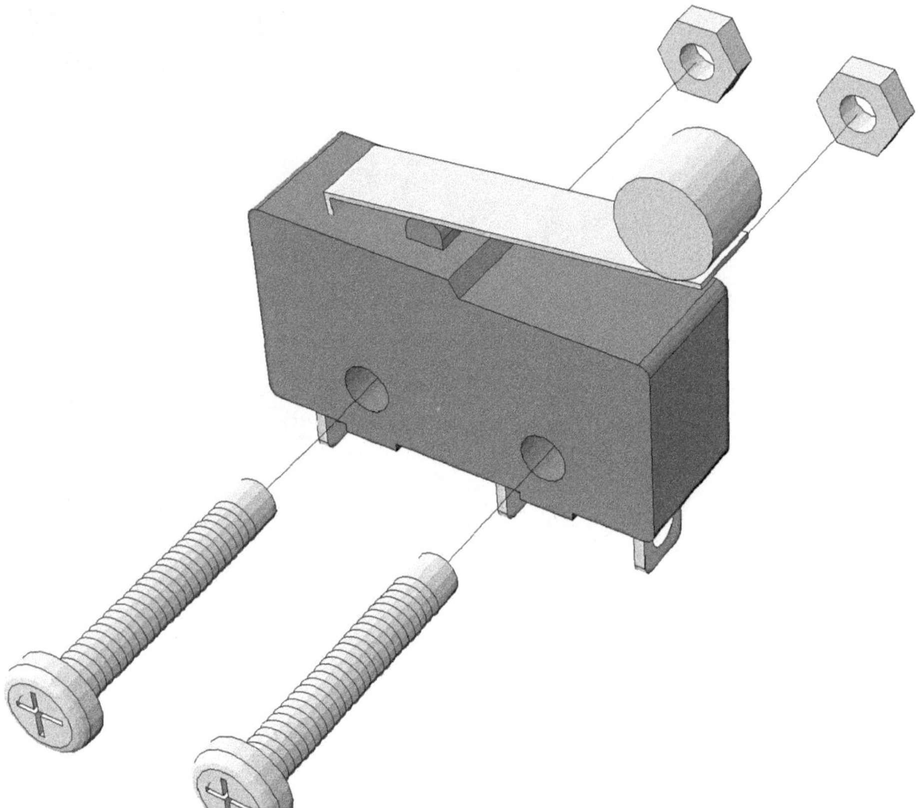

Figure 9.5

Step 6 - Attach the Switch

Slip the ends of the screws and nuts on the switch assembly into the slots of the top plates, as shown in Figure 9.6. Holding the plastic portion of the switch flush with the top plates tighten the two machine screws until the switch is held firmly in place.

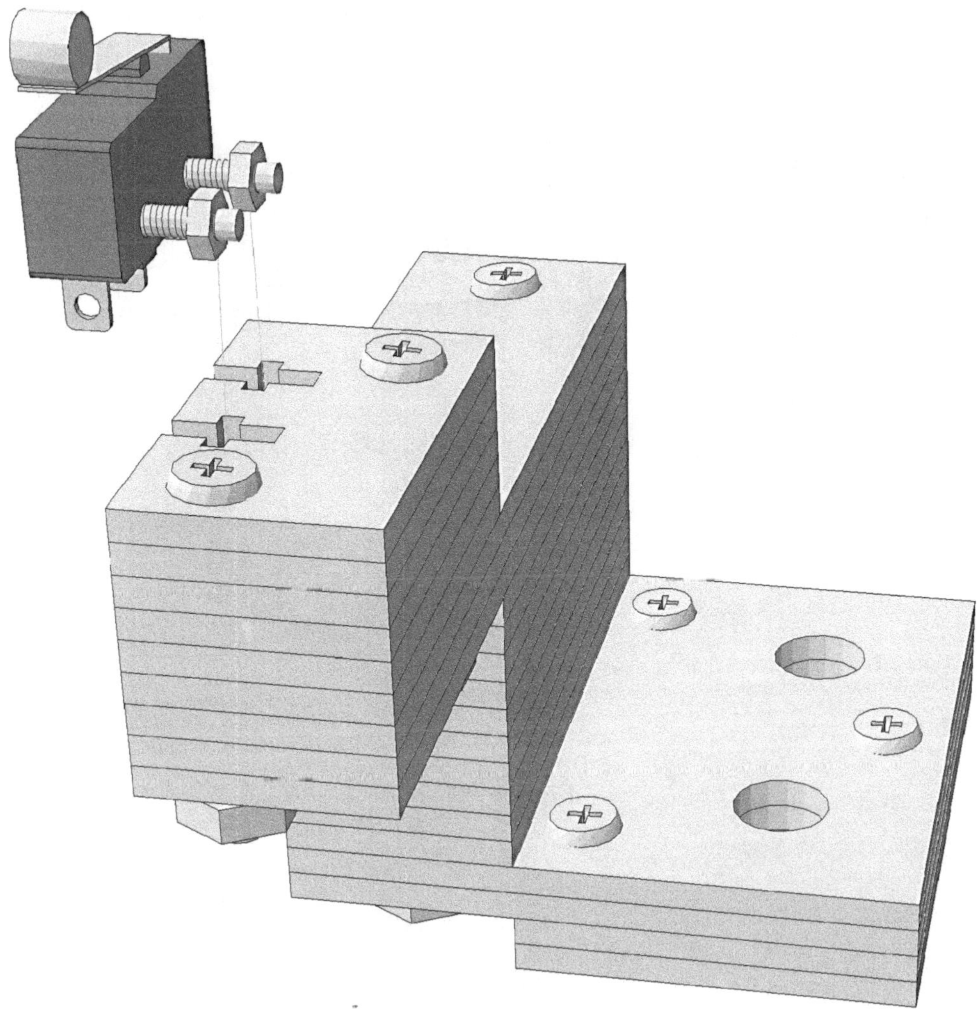

Figure 9.6

The final assembly should look like the one shown in Figure 9.7. Tighten the four machine screws that are attached to nuts. Don't over tighten or you may crack the acrylic. Tighten them enough so that the plates are held together firmly and do not shift.

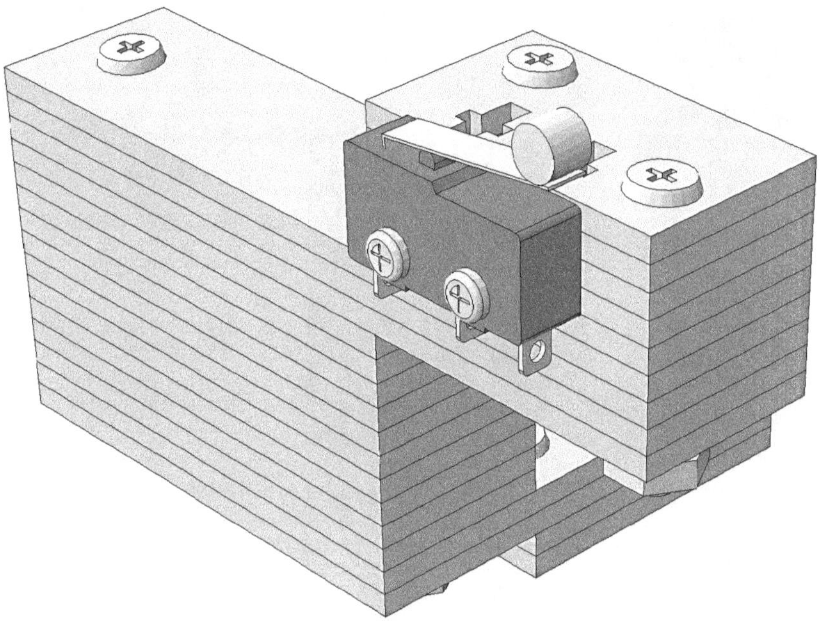

Figure 9.7

Y Axis Homing Switch Installation

Step 1 - Attach Cable

You will need about 12' of cable to attach the Y axis switch to your G540 controller. Strip 1/4" of the insulation from the two conductors on one end of the cable as shown here in Figure 9.8.

Optional: If you like, you can slip two small pieces of heat shrink tubing over the wires before soldering in place. 1/8" tubing works well.

Slip the bare ends of the wire through the two holes in the switch contacts. Notice the switch lead orientation. You want to connect the two wires to the Normally Opened contacts. These are marked NO and C.

Don't worry about the color of the wire connected as polarity does not matter. Solder the wires in place, as shown in Figure 9.8.

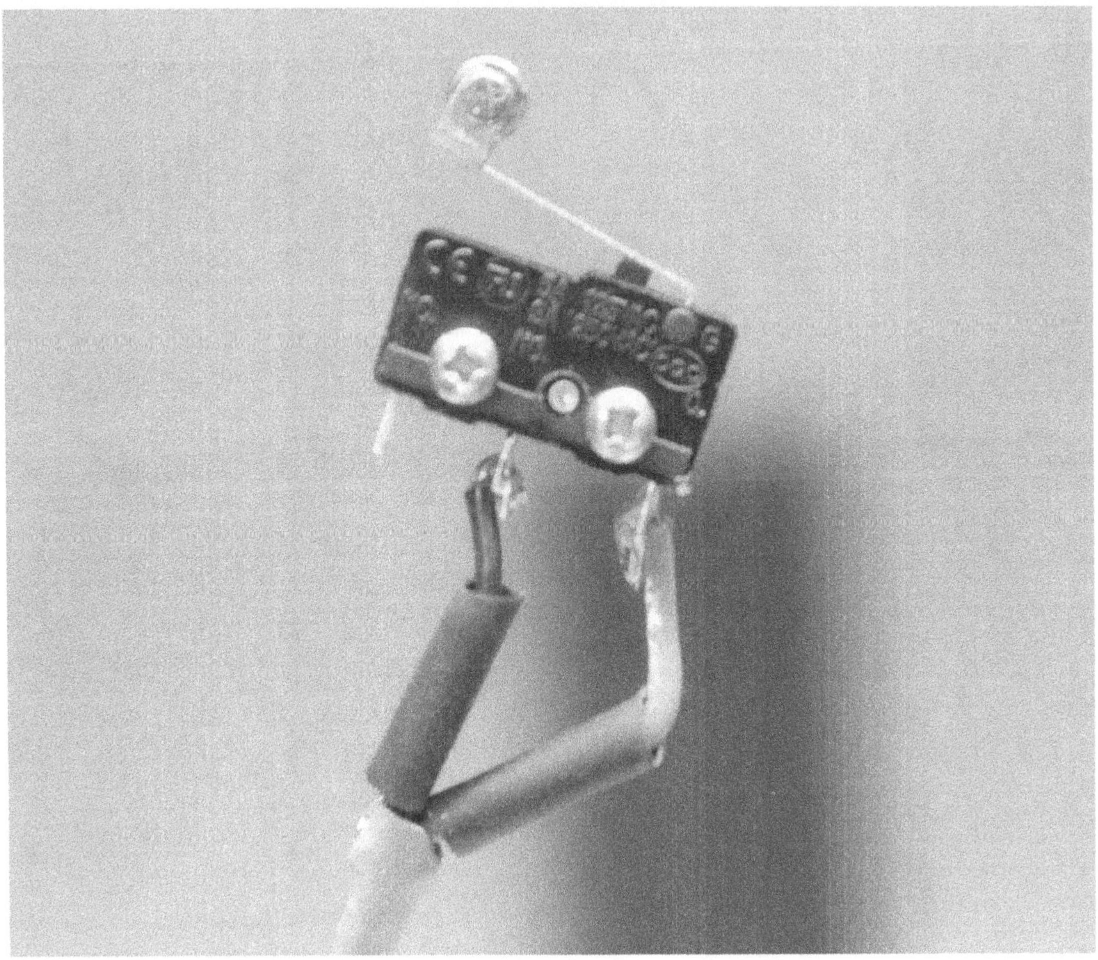

Figure 9.8

Step 2 - Optional

If you did install heat shrink tubing, slide it down over the contacts and apply a small amount of heat until the tubing shrinks around the connectors.

Step 3 - Attach Switch Mount

First place a 1/4" washer on each of the accessory bolts , then slip the switch mount assembly over the two bolts. Place a drop or two of thread lock on the end of each bolt.

Next, add a 1/4" washer and hex nut to each bolt, as shown in Figure 9.9.

Place the top portion of the mount up against the bearing block as shown. Finger tighten the nuts.

Figure 9.9

Step 4 - Attach Switch to Mount

If you have not done so already, attach the switch to the mount as shown in Figure 9.10.

Adjust the switch up and down so that the Y carriage can travel over the roller, slightly depressing it.

Figure 9.10

Conclusion

The Y and A switch assemblies will require more detailed adjustments later.

Figure 9.11

Chapter 10

A Homing Switch

This chapter will take you step-by-step through the assembly of the KRMx01 A homing switch kit assembly.

Tools Needed For This Chapter

- Phillips screwdriver
- Thread lock

Components Needed For This Assembly

- Kronos Robotics KRMx01 A homing switch kit. (see text)

The kit can be purchased from the Kronos Robotics web site here:
http://www.kronosrobotics.com/xcart/product.php?productid=16589

The kit consists of the following components.

- 23, Acrylic pieces
- 2, #6 x 1.5" Machine screws
- 2, #6 x 1" Machine screws
- 3, #6 x 1/4" Machine screws
- 4, #6 Lock washers
- 4 #6 Hex nuts
- 2, #2 x 1/2" Machine screws
- 2, #2 Hex nuts
- 1, Micro switch with roller

Prerequisites

Later when you install the switch assemblies on your KRMx01 CNC, it is assumed that you added the extra 1/4-20 x 3/4" carriage bolts as outlined in the first KRMx01 book. These bolts will be used to secure your switch assembly. If you did not add these bolts, you will have to perform some disassembly to install the bolts.

Before starting the kit assembly be sure to remove the protective plastic from both sides of each of the acrylic pieces. Be sure all the holes are clear of debris that may still be present. The easiest way to do this is with the tip of a pencil or pen.

Warning

Acrylic is extremely brittle and until your switch assembly is assembled and installed there is a chance of breakage.

Tip

If you decide to create your own mounts for this homing switch, the wiring and configuration chapters can still be used to help you get them operational. You can purchase just the roller switch and screws by purchasing the Kronos Robotics Roller Switch kit.

http://www.kronosrobotics.com/xcart/product.php?productid=16612

Step 1 - Assemble Base

Sandwich the four acrylic components shown in Figure 10.1 together. Note that the component on the bottom has smaller threaded holes.

Insert the three #6 x 1/4" machine screws and tighten until they are snug only. They are only used to align the four pieces until they are attached to the CNC.

Place a small amount of thread lock on the tips of the three screws before inserting.

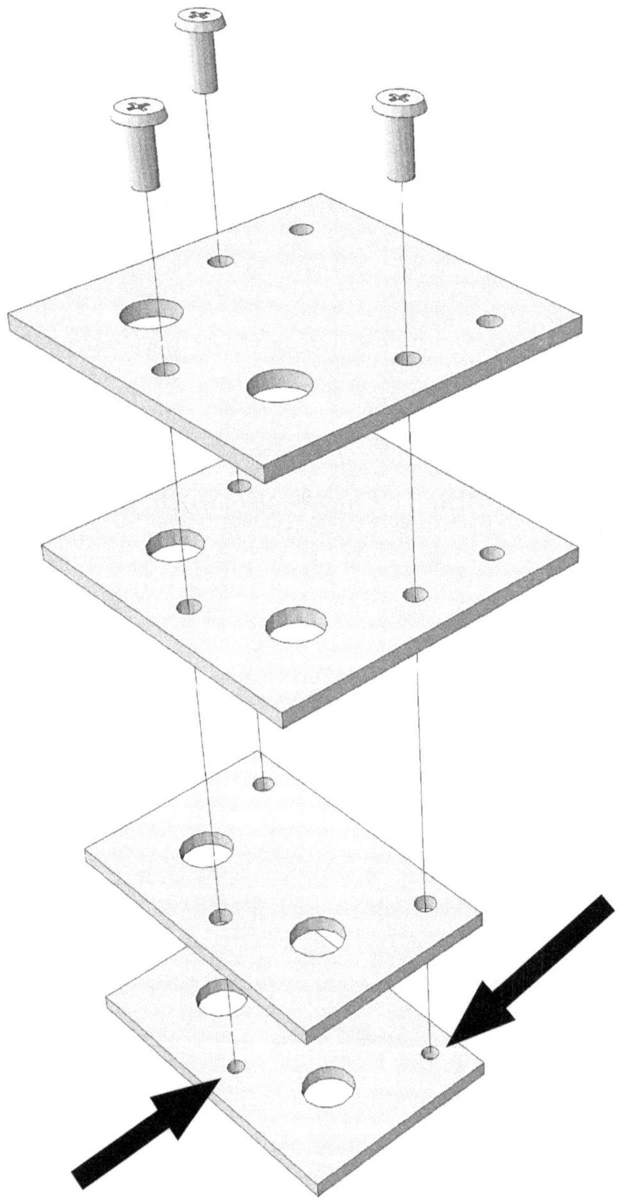

Figure 10.1

Step 2 - Attach Riser to Base

Take the three longer riser plates and insert the two #6 x 1-1/2" machine screws into the two center holes. Notice the orientation of the two outer holes, as shown in Figure 10.2.

Next, add ten of the smaller riser plates to the machine screws.

Insert the riser assembly into the base holes, again noting the orientation.

Add a #6 lock washer and nut to each machine screw and finger tighten.

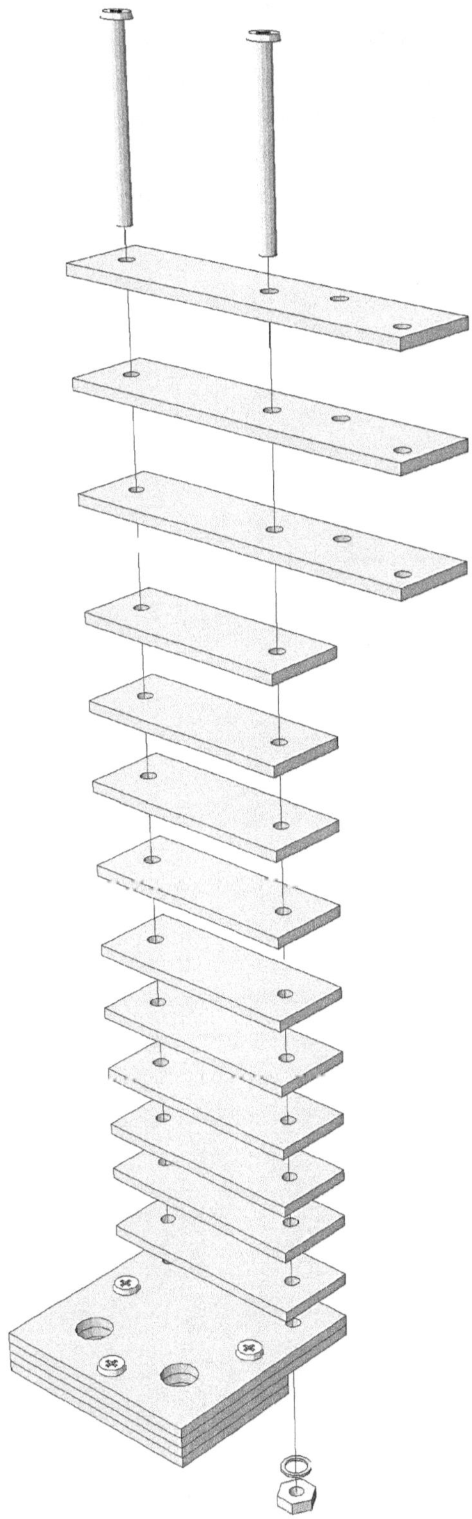

Figure 10.2

Step 3 - Assemble Top

Sandwich the six acrylic components shown in Figure 10.3 together. Notice the orientation of the holes.

Insert the two #6 x 1" machine screws into the two holes shown in Figure 10.3.

Don't attach any nuts at this time.

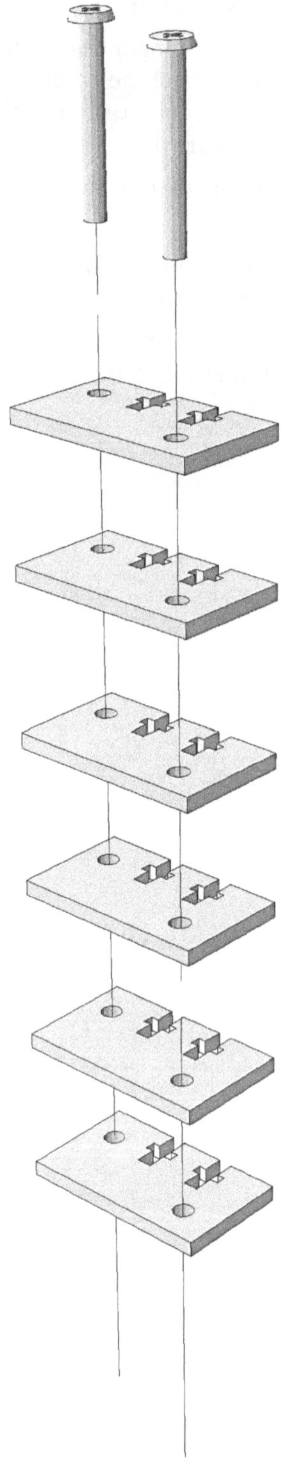

Figure 10.3

Step 4 - Attach Top to Assembly

Insert the two #6 x 1" machine screws into the two matching holes on the long risers.

Add a #6 lock washer and hex nut and finger tighten.

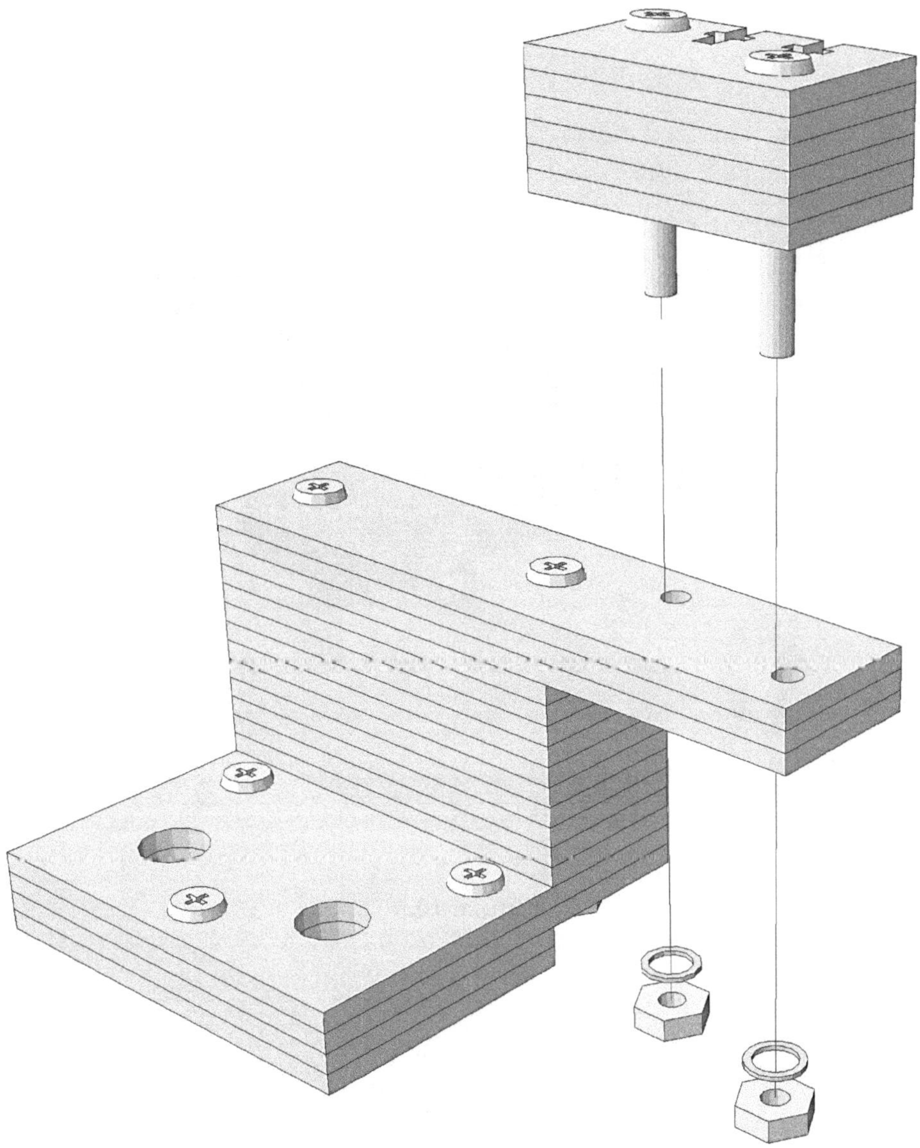

Figure 10.4

Step 5 - Assemble the Switch

Insert two #2 x 1/2" machine screws into the two holes on the micro switch, as shown in Figure 10.5. Pay attention to the orientation of the roller on the switch.

Add two #2 hex nuts to the ends of the screws. Do not tighten the nuts, leave them on the ends of the screws.

Add a couple drops of thread lock to the nuts.

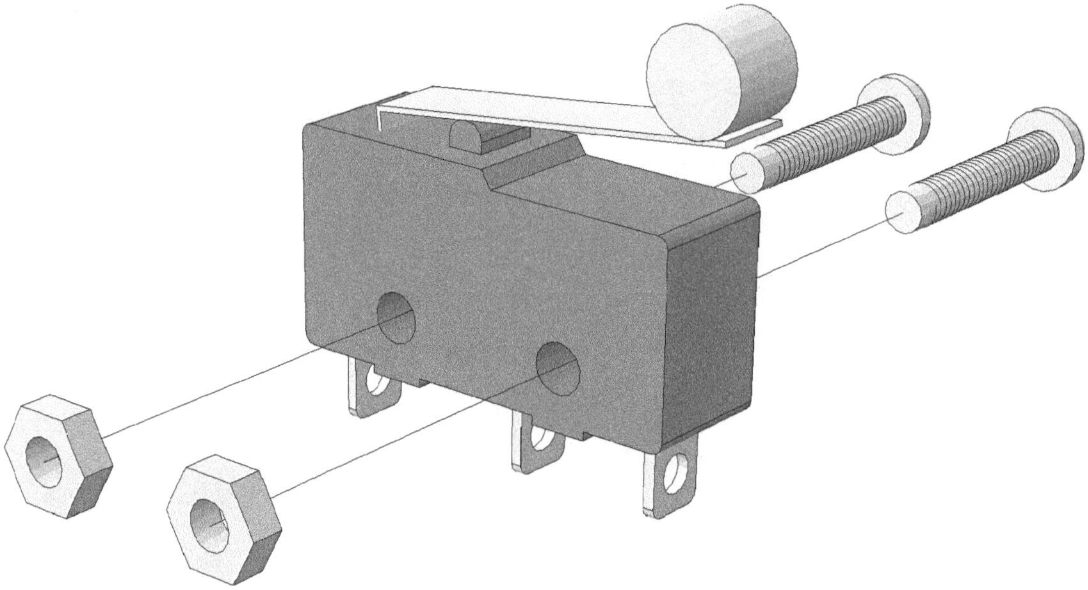

Figure 10.5

Step 6 - Attach the Switch

Slip the ends of the screws and nuts on the switch assembly into the slots of the top plates, as shown in Figure 10.6. Holding the plastic portion of the switch flush with the top plates tighten the two machine screws until the switch is held firmly in place.

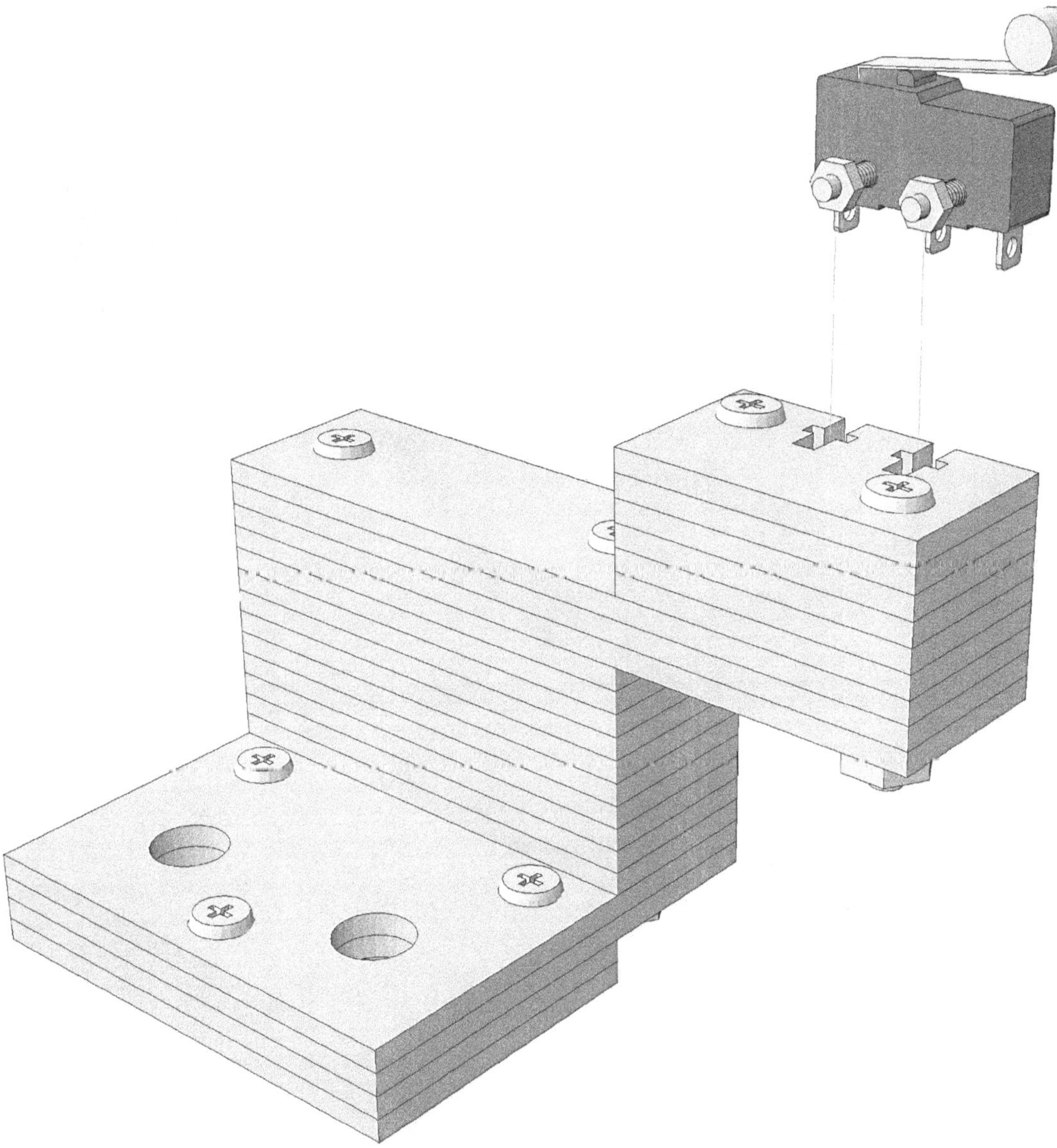

Figure 10.6

The final assembly should look like the one shown in Figure 10.7. Tighten the four machine screws that are attached to nuts. Don't over tighten or you may crack the acrylic. Tighten them enough so that the plates are held together firmly and do not shift.

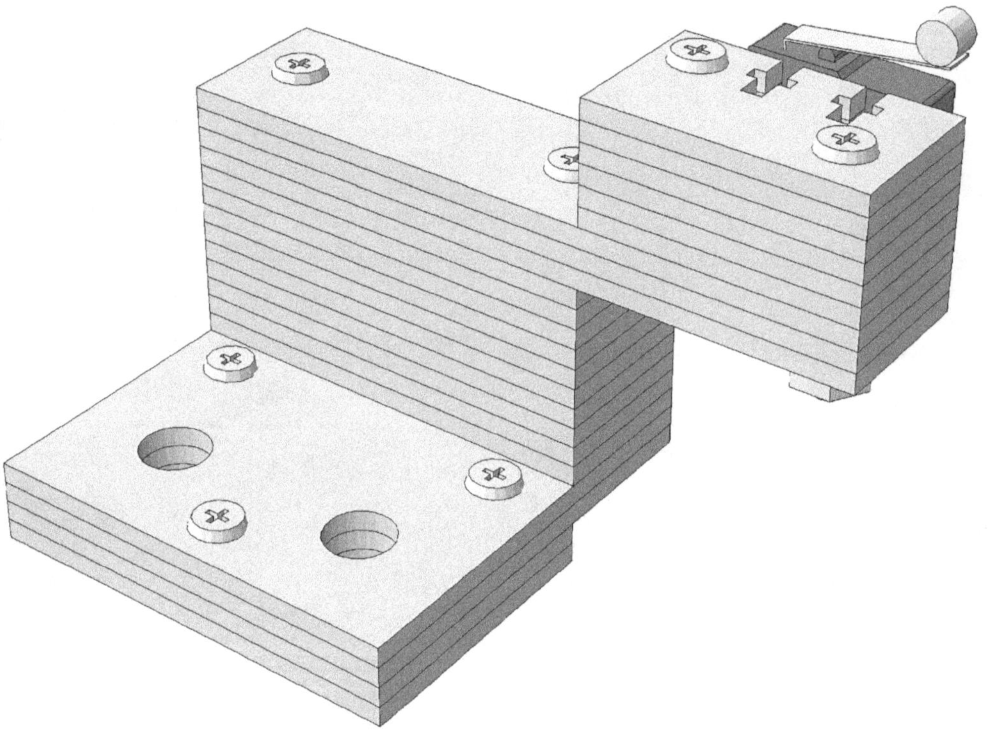

Figure 10.7

A Axis Homing Switch Installation

Step 1 - Attach Cable

You will need about 12' of cable to attach the A axis switch to your G540 controller. Strip 1/4" of the insulation from the two conductors on one end of the cable.

Optional: If you like, you can slip two small pieces of heat shrink tubing over the wires before soldering in place. 1/8" tubing works well.

Slip the bare ends of the wire through the two holes in the switch contacts. Notice the switch lead orientation. You want to connect the two wires to the Normally Opened contacts. These are marked NO and C.

Don't worry about the color of the wire connected as polarity does not matter. Solder the wires in place, as shown in Figure 10.8.

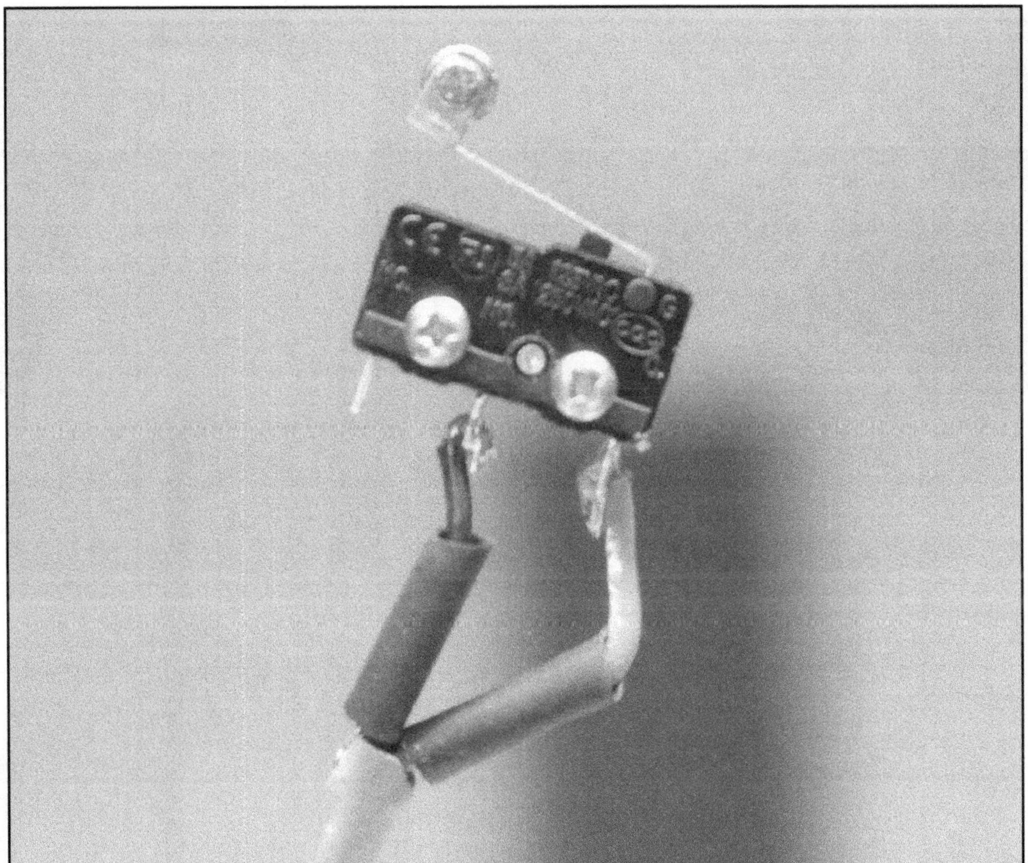

Figure 10.8

Step 2 - Optional

If you did install heat shrink tubing, slide it down over the contacts and apply a small amount of heat until the tubing shrinks around the connectors, as shown in Figure 10.2.

Step 3 - Attach Switch Mount

First place a 1/4" washer on each of the accessory bolts , then slip the switch mount assembly over the two bolts. Place a drop or two of thread lock on the end of each bolt.

Next, add a 1/4" washer and hex nut to each bolt, as shown in Figure 10.9.

Place the top portion of the mount up against the bearing block as shown. Finger tighten the nuts.

Figure 10.9

Step 4 - Attach Switch to Mount

If you have not done so already, attach the switch to the mount as shown in Figure 10.10.

Adjust the switch up and down so that the Y carriage can travel over the roller, slightly depressing it.

Figure 10.10

Conclusion

The addition of the A switch when used in conjunction with the Y switch, will allow you to square your gantry to the table. The Y and A switch assemblies will require more detailed adjustments later.

Figure 10.5

Chapter 11

Z Homing Switch

The KRMx01 Z axis switch does not have a mounting assembly. It mounts directly to the Z bearing brackets using two #2 machine screws and nuts.

Tools Needed For This Chapter

- Small Phillips screwdriver
- 3/32" Drill bit
- Drill
- Soldering iron and solder
- Thread lock

Components Needed For This Assembly

- Kronos Robotics KRMx01 Z homing switch kit. (see text)

The kit can be purchased from the Kronos Robotics web site here:
http://www.kronosrobotics.com/xcart/product.php?productid=16592

The kit consists of the following components.

- 2, #2 x 1/2" Machine screws
- 2, #2 Hex nuts
- 1, Micro switch with roller

Prerequisites

None

Step 1 - Attach Cable

You will need about 15' of cable to attach the Z axis switch to your G540 controller. Strip 1/4" of the insulation from the two conductors on one end of the cable.

Optional: If you like you can slip two small pieces of heat shrink tubing over the wires before soldering in place. 1/8" tubing works well.

Slip the bare ends of the wire through the two holes in the switch contacts. Notice the switch lead orientation. You want to connect the two wires to the Normally Opened contacts. These are marked NO and C.

Don't worry about the color of the wire connected as polarity does not matter. Solder the wires in place, as shown in Figure 11.8.

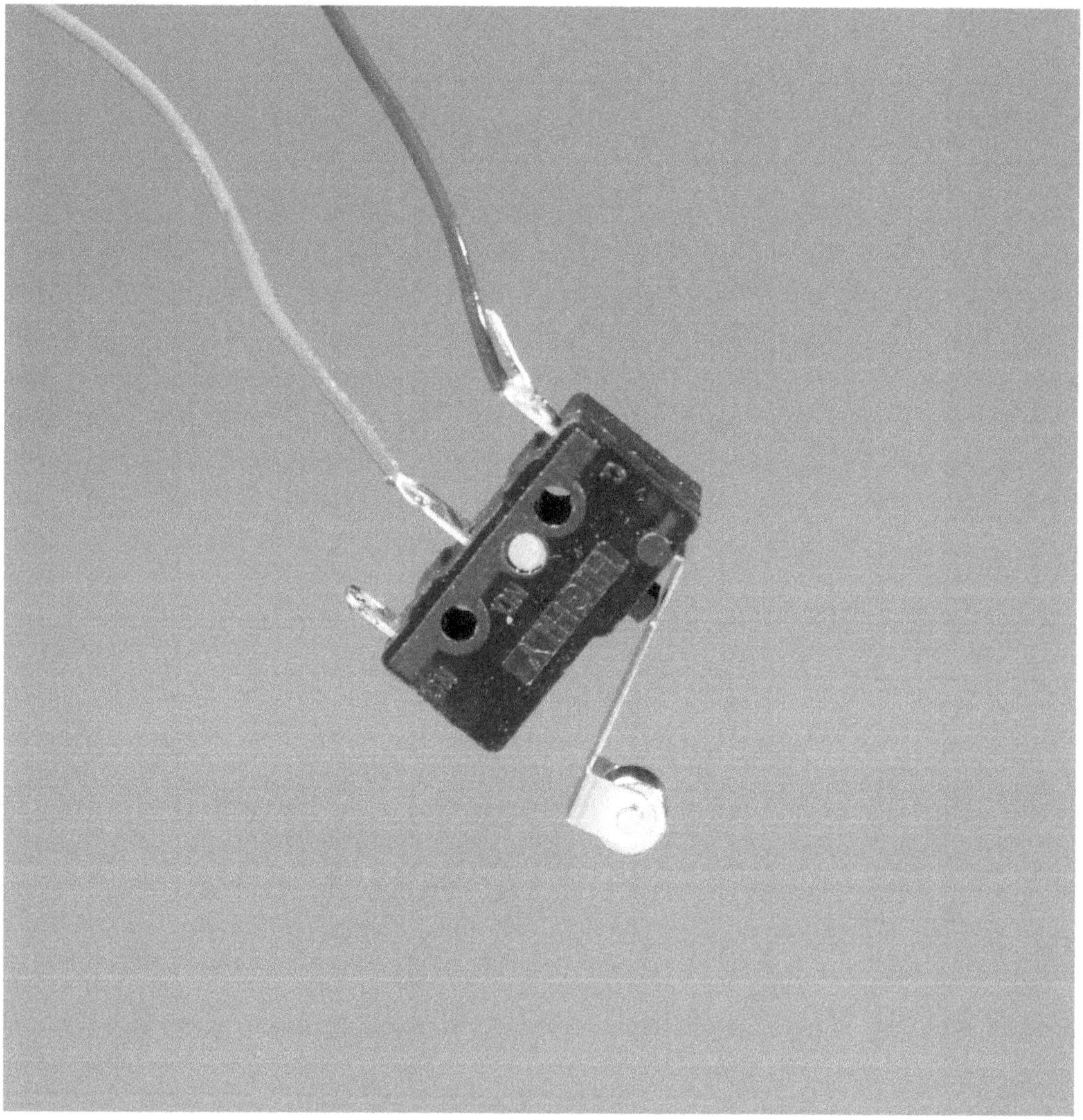

Figure 11.1

Step 2 - Optional

If you did install heat shrink tubing, slide it down over the contacts and apply a small amount of heat until the tubing shrinks around the connectors, as shown in Figure 11.2.

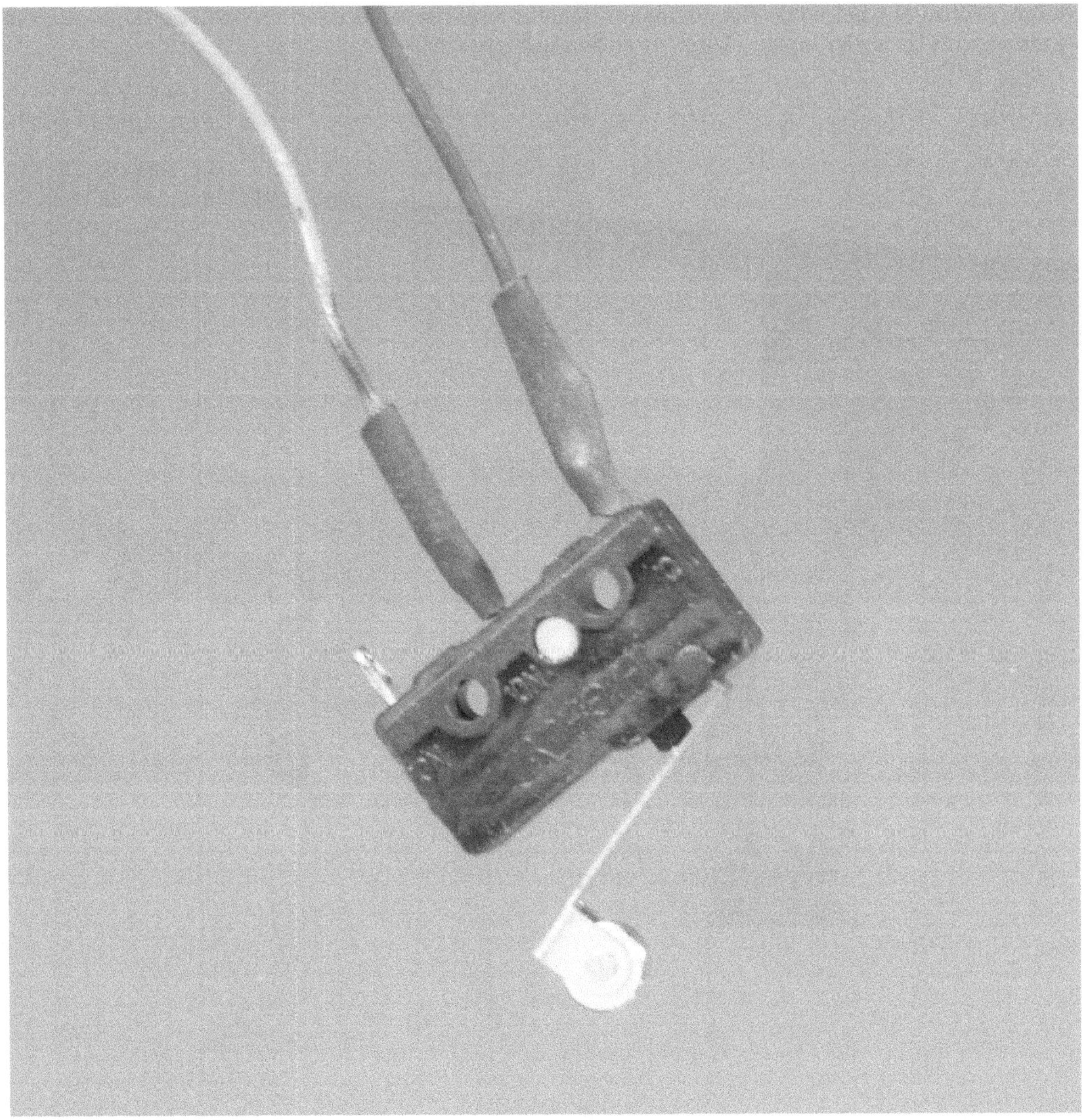

Figure 11.2

Step 3 - Drill Mounting Holes

Hold the switch up against the rail as shown in Figure 11.3. Using a 3/32" drill bit, drill the first hole into the bearing bracket. Once the hole is drilled, insert one of the #2 machine screws to hold the switch in place and drill the second hole.

!!Important!!

Before drilling the holes test the alignment to make sure the top bracket on the Z carriage will come in contact with the roller on the switch as shown in Figure 11.4.

Figure 11.3

Figure 11.4

Conclusion

It is possible for the carriage to damage the switch if the roller is driven into the switch. If yours does, you can add a small bolt and washer to one of the holes in the top bracket of the Z carriage to act as a stop, as shown in Figure 11.5.

Figure 11.5

Chapter 12

Homing Switch Wiring

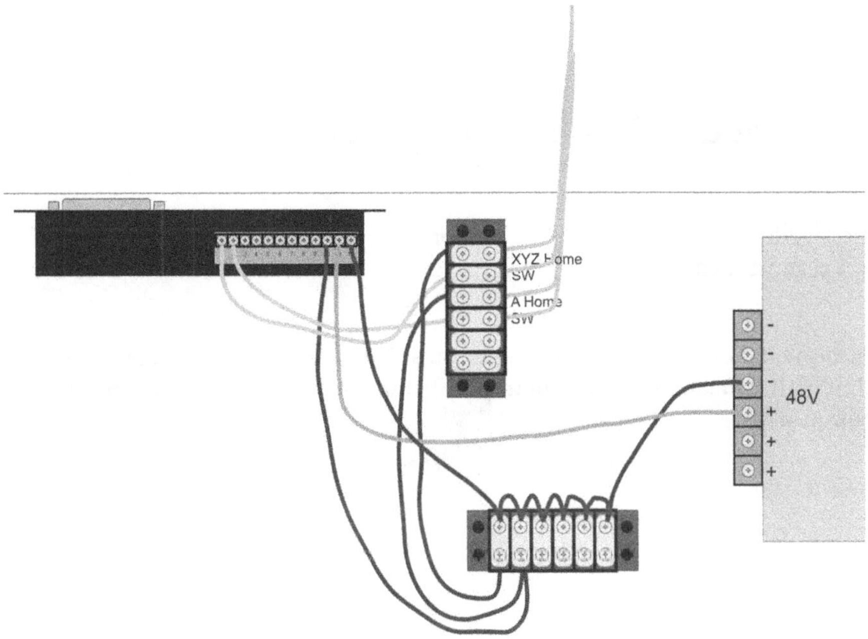

With the switches mounted, it's time to rout the cables and hook them up to the G540 controller on the KRMx01 CNC.

Tools Needed For This Chapter

- Phillips screwdriver
- Wire cutters

Components Needed For This Chapter

- 4, Small tie wraps
- 2, Twist on terminal caps

The cable can be just about any 16-24 gauge wire. Stranded will be easier to work with but solid will work if that's all you can get. Cut about 6" off this cable. Later in the chapter you will use cutoff portion as hookup wire.

Prerequisites

The wiring shown here assumes you have already installed and wired your emergency stop switch. The emergency stop switch project installs an additional terminal strip (shown in Figure 12.1) that will be used in this project.

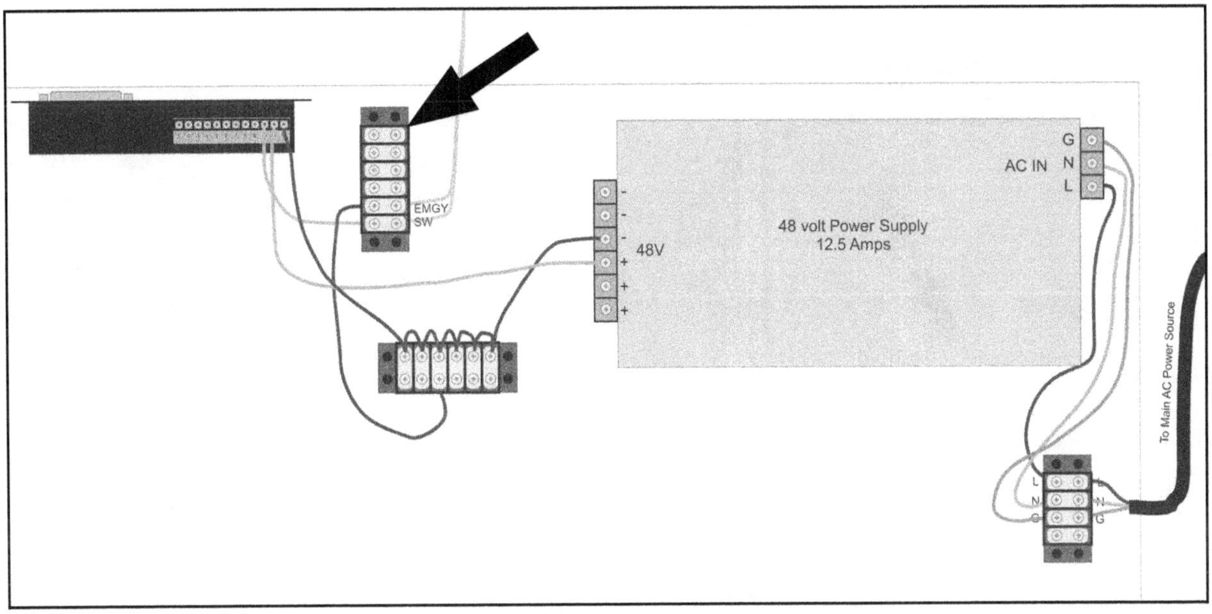

Figure 12.1

Step 1 - Wire XYZ Terminals

The X, Y and Z switches are all connected together and in turn connected between terminal 1 on the G540 and ground. To facilitate this we will use the first two terminals on the upper terminal block.

Take a piece of hookup wire and connect it between terminal 1 on the G540 the second terminal on the upper terminal block, as shown in Figure 12.2 Connect another wire between the two terminal blocks as shown.

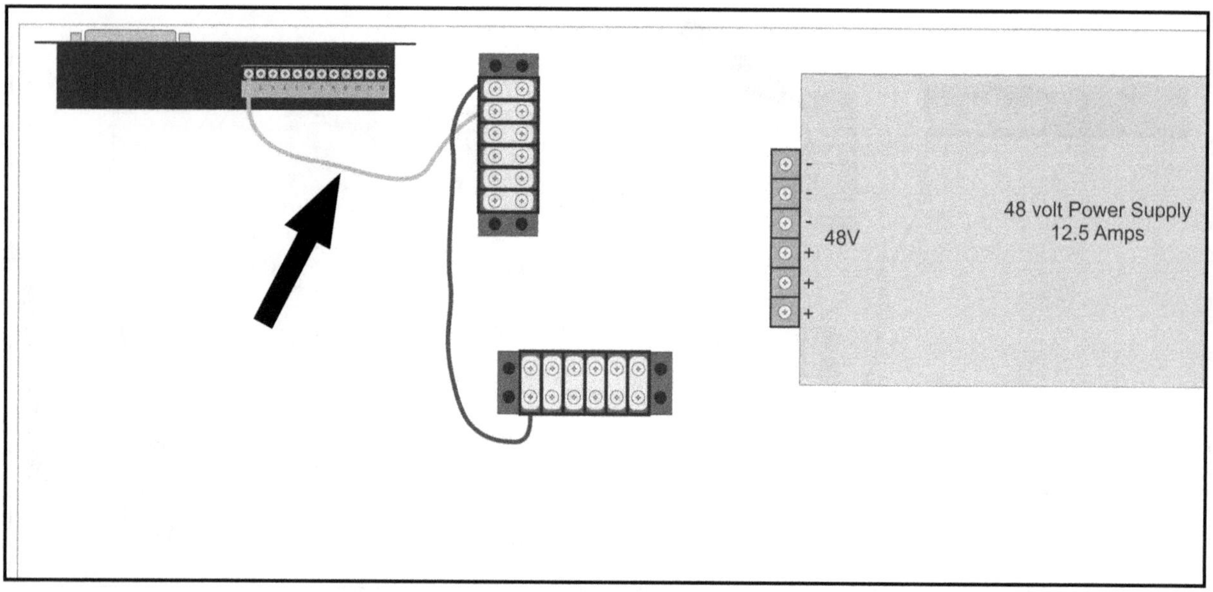

Figure 12.2

Step 2 - Wire A Terminals

The A switch is connected between terminal 2 on the G540 and ground. To facilitate this we will use the third and fourth terminals on the upper terminal block.

Take a piece of hookup wire and connect it between terminal 2 on the G540 the third terminal on the upper terminal block, as shown in Figure 12.3 Connect another wire between the two terminal blocks as shown.

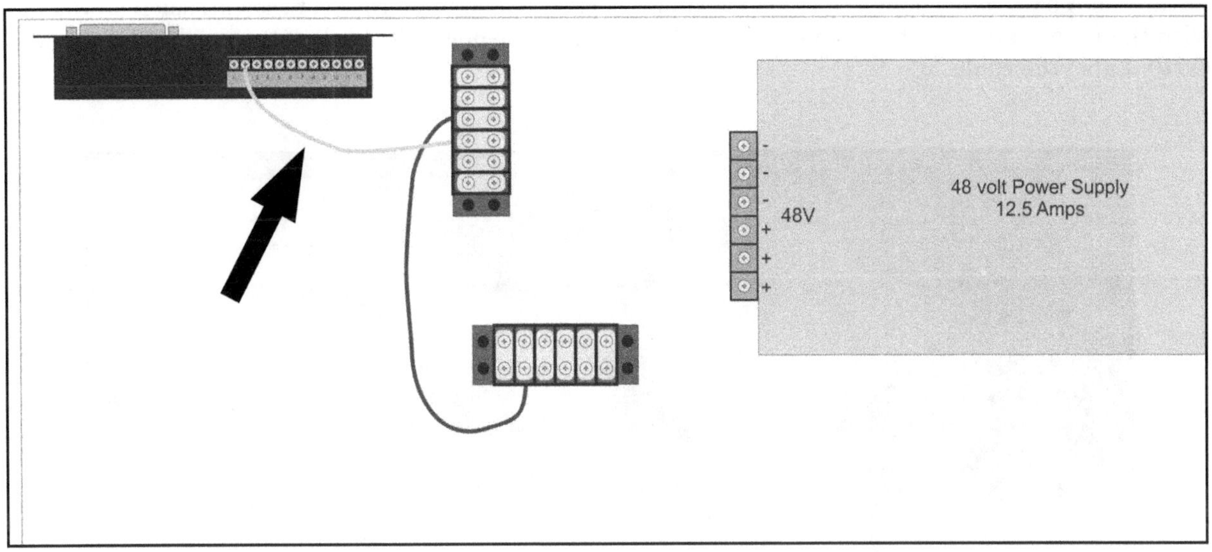

Figure 12.3

Step 3 - Rout the cables

Run the Z axis cable up the top of the Z motor block and down the same path as the Z motor cable. This should take you through the X dragon cable then through the Y dragon cable then down the back of the CNC. Run the cable to the location of your G540 controller. Label the cable.

Run the X axis cable through the Y dragon cable and down the back of the CNC. Again, run the cable to the location of G540 and label the cable.

Run the Y axis cable along the left Y beam. Insert the cable into groves of the extrusion as you did with the Estop cable, as shown in Figure 12.4. Run the cable down the back of the machine and to the G540. Label the cable.

Run the A axis cable along the right Y beam. Insert the cable into groves of the extrusion as you did with the Estop cable, as shown in Figure 12.4. Run the cable down the back of the machine and to the G540. Label the cable.

Figure 12.4

Step 4 - Group the Cables

Take all the cables and tie them together at various points until you reach the G540. Giving yourself a little extra room place a tie wrap on the cable near where the G540 is located. You have a choice of cutting off the excess cable or winding it up and tieing it together with tie wraps. The goal is to get all the cables together at the end.

Expose the conductors on each cable about 3" then strip about 1" of the insulation from each of the cables.

Take one lead from each X, Y, and Z cable and twist them together, along with 5" piece of hookup wire. Secure them with a terminal cap as shown in Figure 12.5.

Do the same with the second lead from each X, Y, and Z cable. On they Y cable expose the conductors and strip 1/2" from each of the conductors.

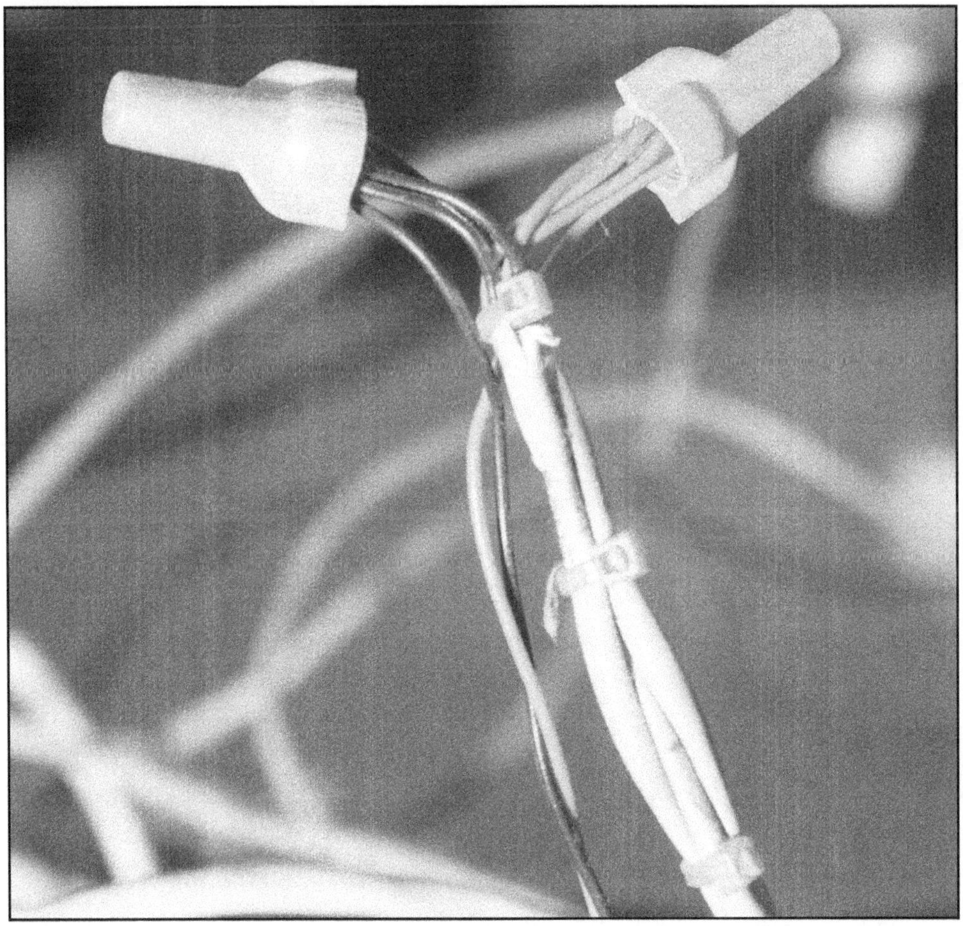

Figure 12.5

Step 5 - Connect X, Y and Z to Terminals

Connect the small hookup wires you added to the X, Y and Z cables to the first two terminals on the upper terminal block as shown in Figure 12.6.

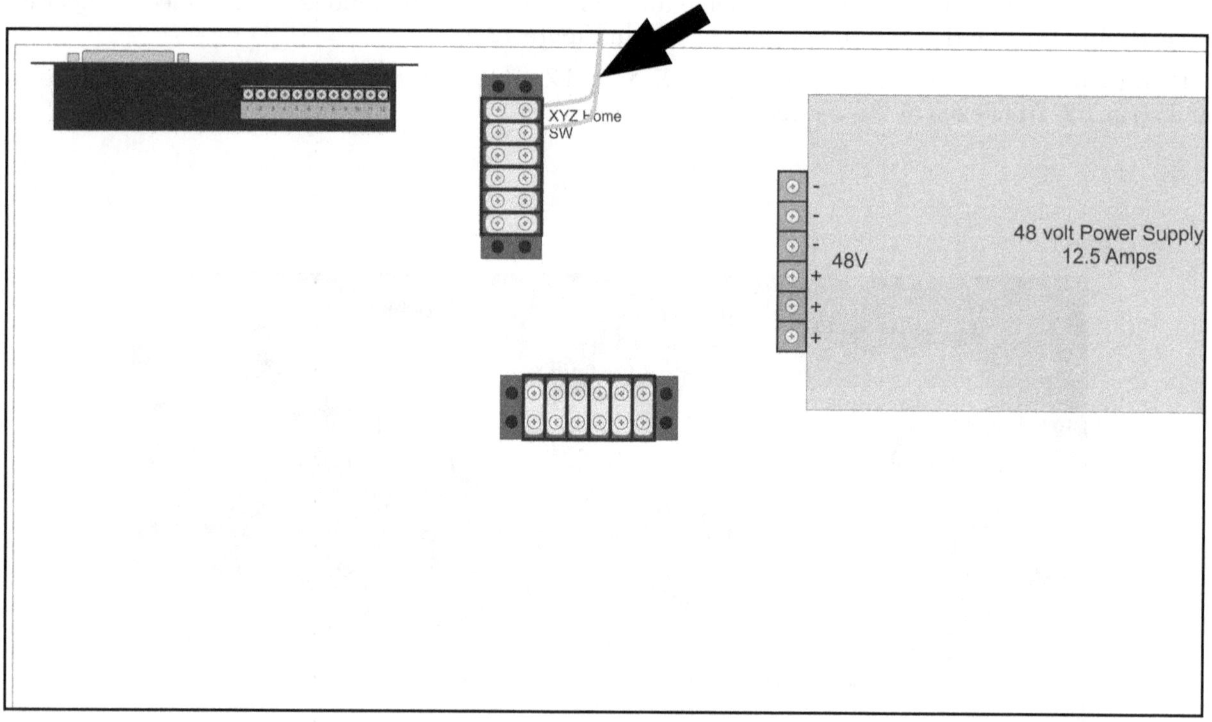

Figure 12.6

Step 6 - Connect A Terminals

Connect the A switch cable to the third and fourth terminals on the upper terminal block as shown in Figure 12.7.

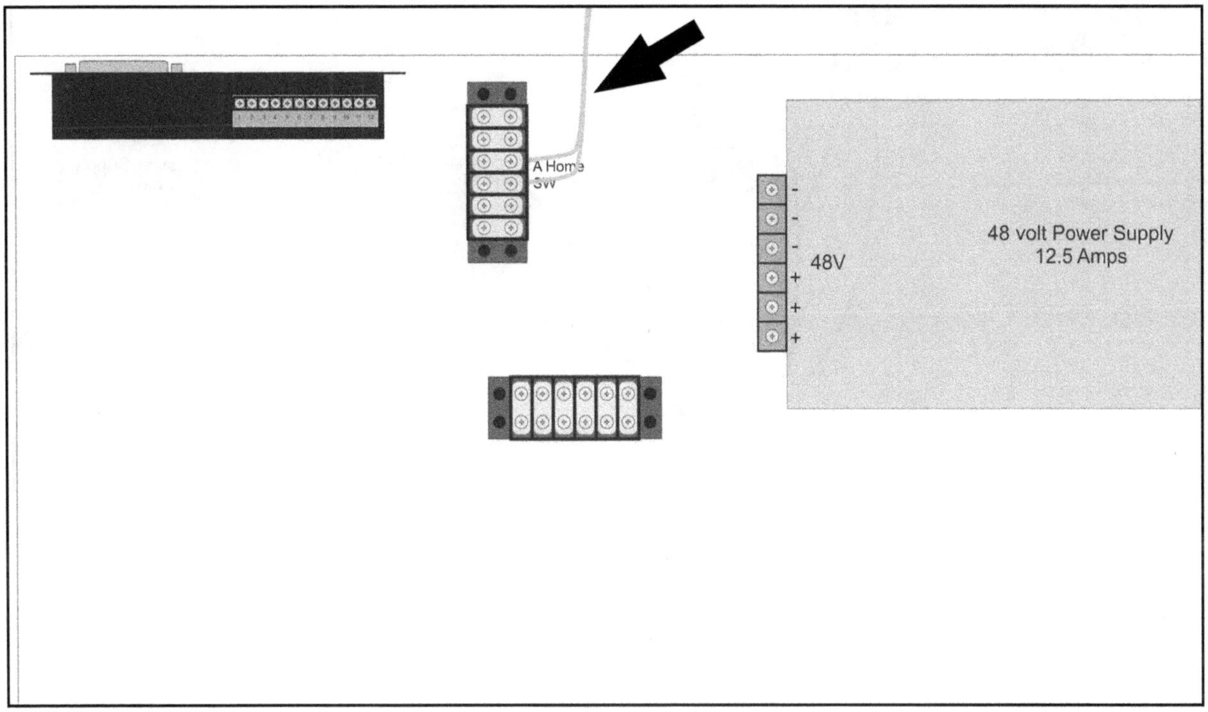

Figure 12.7

Conclusion

The complete wiring layout is shown in Figure 12.8. Double check it. Note that the Estop cable is not shown.

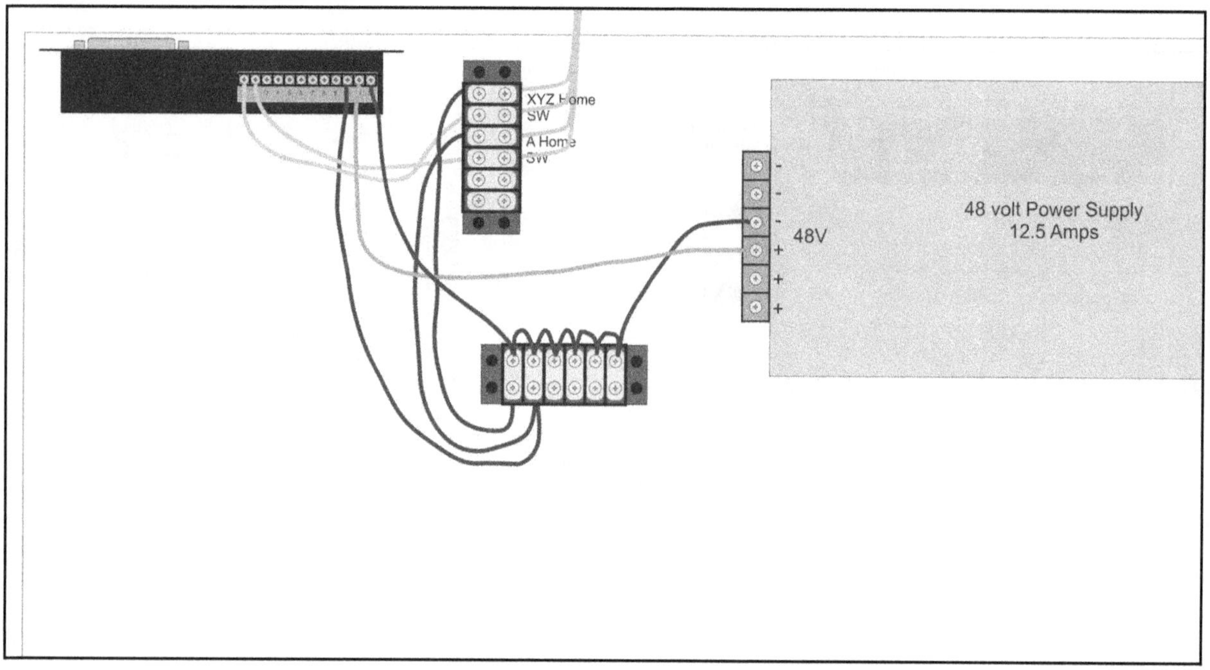

Figure 12.8

Chapter 13

Homing Switch Configuration

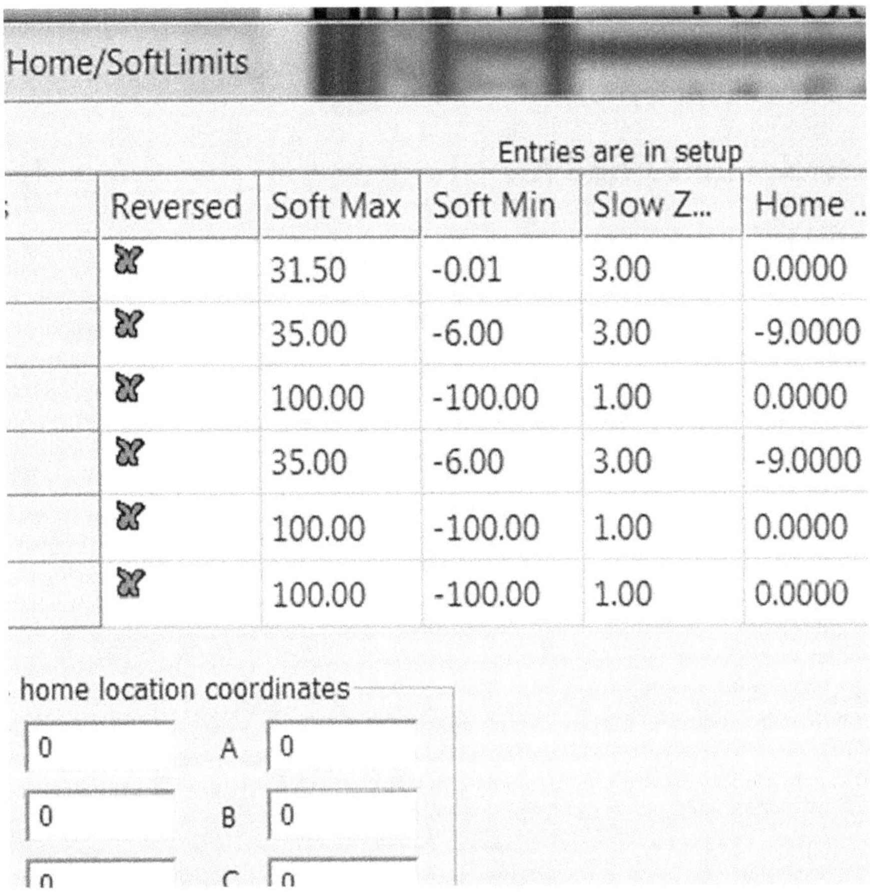

With the all the switches installed and wired, it's time to let Mach3 know how to use them.

Tools Needed For This Chapter

None

Components Needed For This Chapter

None

Prerequisites

The switches should be mounted and all the wiring in place.

Step 1 - Configure Homing Options

Enter the Homing/Limits option from the Config menu.

Change the X, Y, Z, and A settings to match the ones shown in Figure 13.1, then hit OK.

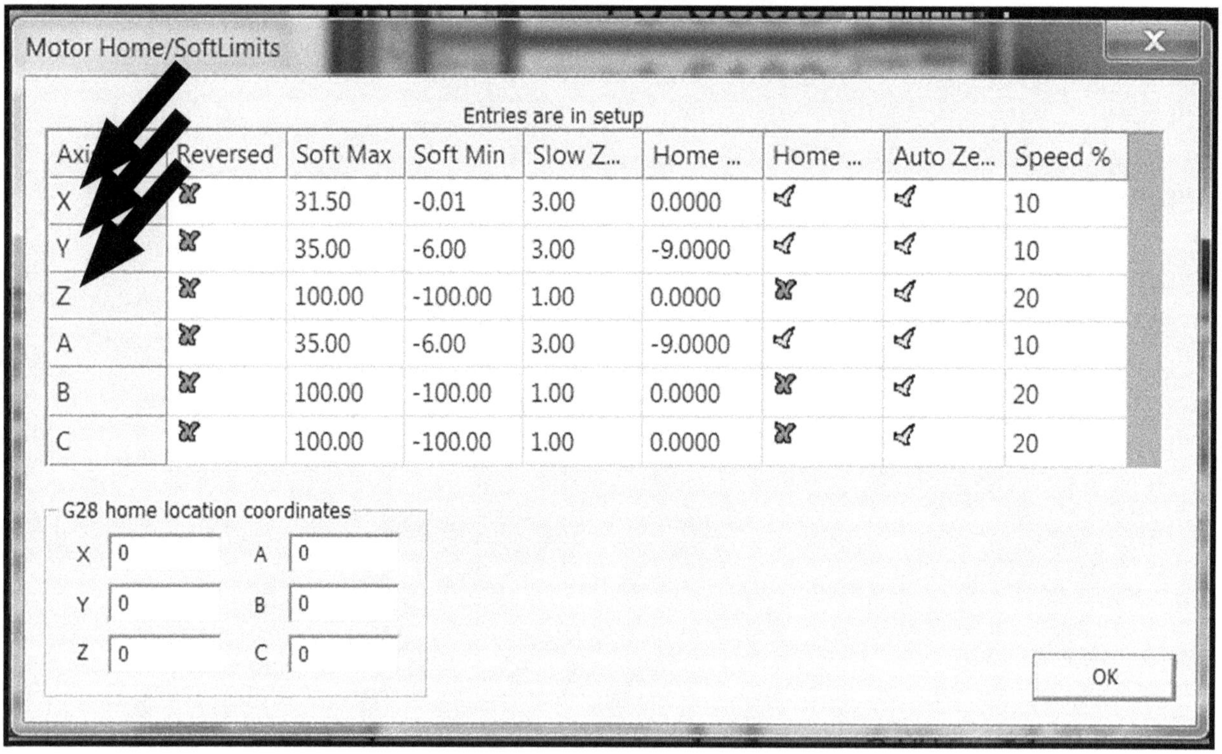

Figure 13.1

Step 2 - Configure X and Z Input Pins

Enter the Ports and Pins option from the Config menu.

Change the X Home settings to match those shown in Figure 13.2. Change the Z Home settings to match those shown in Figure 13.3. Don't change anything else. Hit the OK button when finished.

Figure 13.2

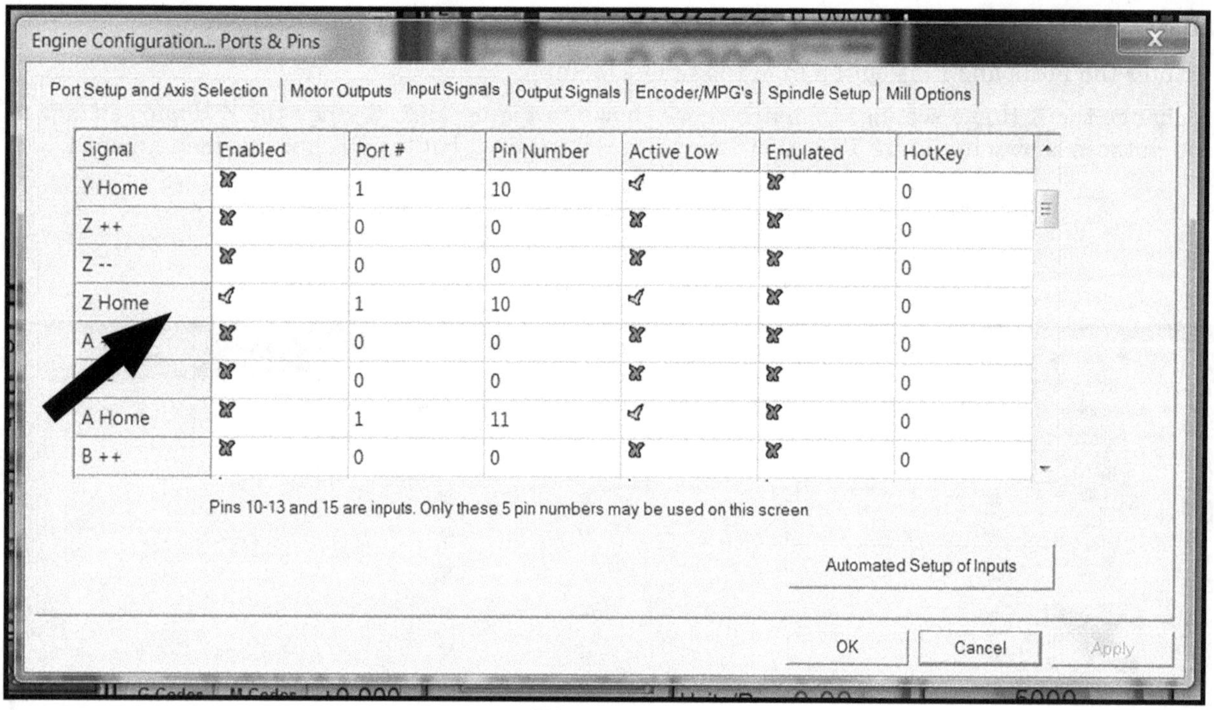

Figure 13.3

Step 3 - Test the X and Z Switches

Move the CNC so that none of the homing switches are being activated (depressed). Press the REF ALL HOME button shown in Figure 13.4.

The CNC should move up until the Z switch is hit then to the left until the X switch is hit. The CNC should then stop.

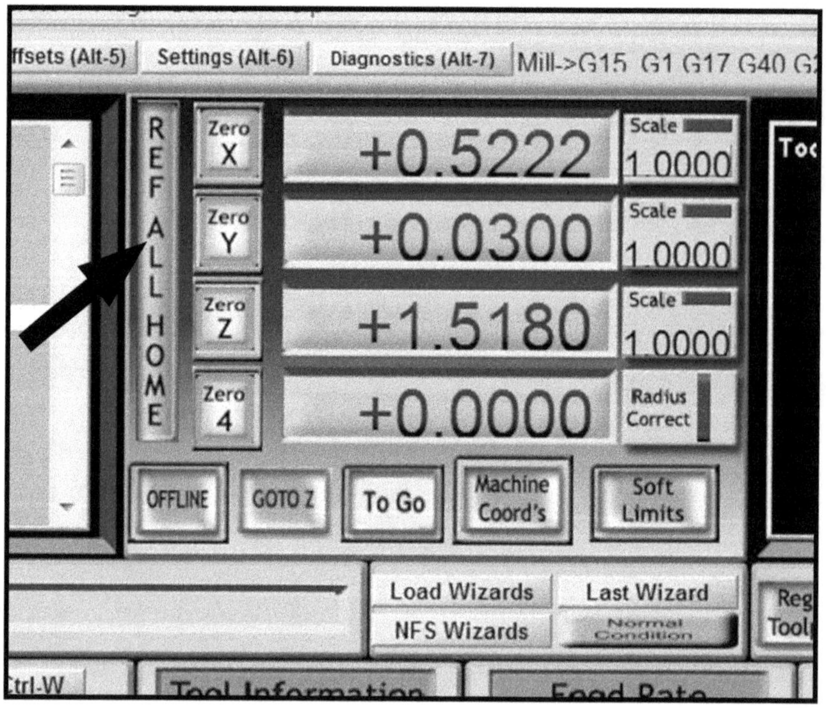

Figure 13.4

If the CNC does not perform as indicated in Step 3, go back and check your wiring. Make sure you wired the normally open contacts on the switches. If the wiring looks good recheck your configuration settings.

In addition, on the on the Diagnostics page there are four small lights shown in Figure 13.5. They are labeled M1Home, M2Home, M3Home, and M4Home. They correspond to the X, Y, Z, and A switches. As long as the switch has been enabled the light will light when the switch is depressed. In this test when you press the either X or Z switch both the M1Home and M3Home lights will light. Its a quick way to check your wiring. If they do light when depressed and the CNC does not behave properly, you probably have a configuration error.

Note that you can manually jog your machine using the arrow keys, while you watch the indicator lights. This way you can determine if the carriages are actually activating the switch. If not you will need to go back and adjust the switches position.

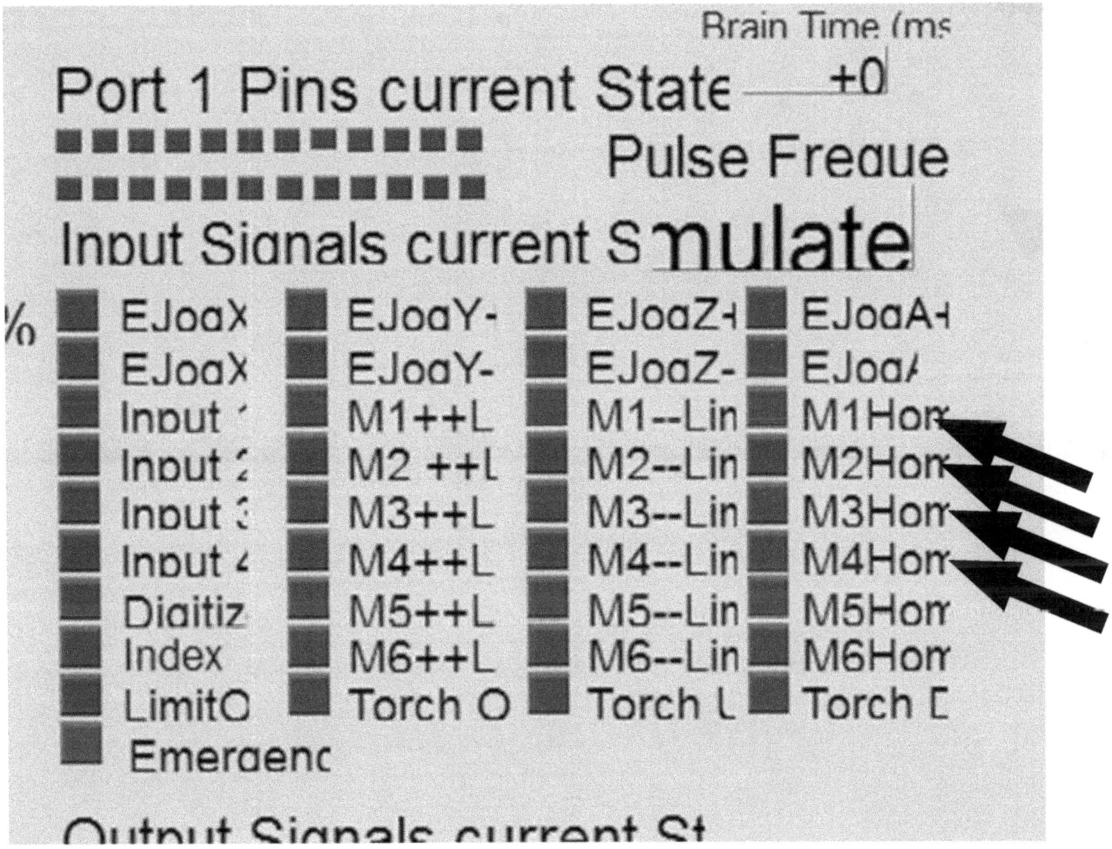

Figure 13.5

Step 4 - Configure the Y and A Input Pins

Enter the Ports and Pins option from the Config menu.

Change the Y Home and A Home settings to match those shown in Figure 13.6. Hit the OK button when finished.

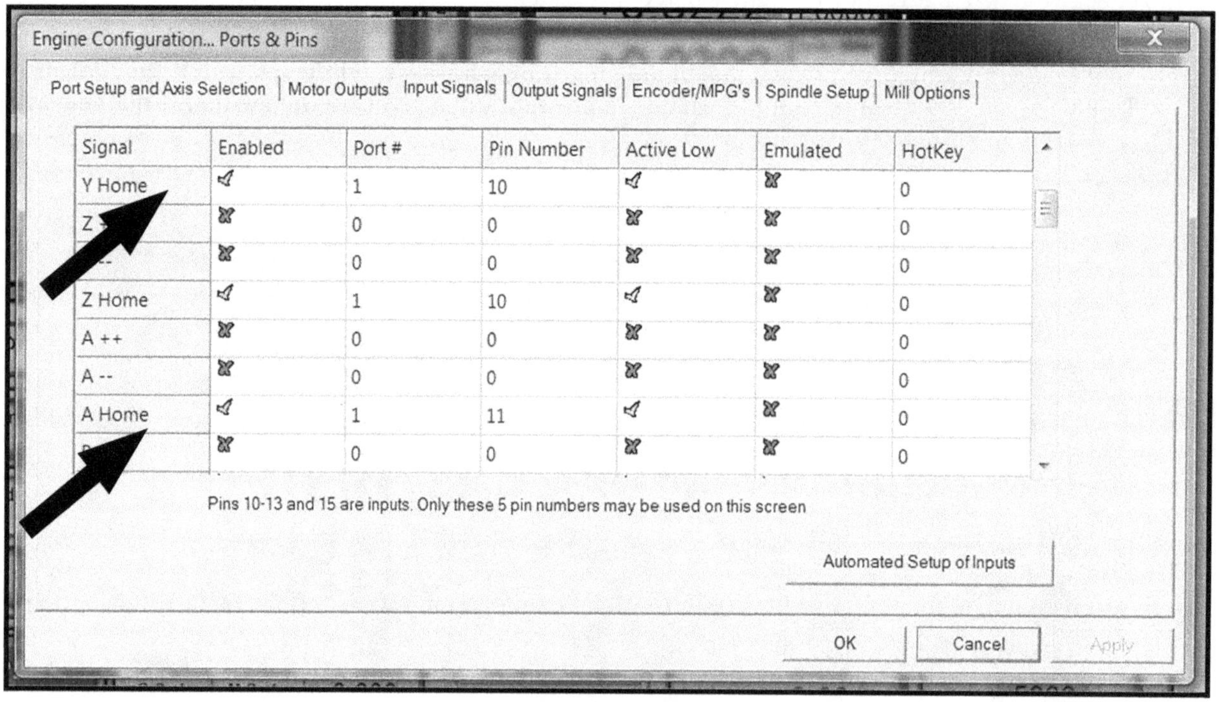

Figure 13.6

Step 5 - Test the X, Y, Z and A Switches

Move the CNC so that none of the homing switches are being activated. Press the REF ALL HOME button.

The CNC should move up until the Z switch is hit, then the gantry will move forward until the Y switch is hit. The X carriage will move left until the X switch is hit. Then the A side of the gantry will move in and out until the A switch is cycled. The CNC should then stop.

If it does not, be ready to hit your Estop button to shut down the motors. Check the Diagnostics page against all switches. If you press any of the X, Y or Z switches all the M1Home, M2Home, and M3Home lights should illuminate.

Pressing the A switch should illuminate the M4Home switch. If all the switches are light the appropriate lights then the problem is in your configuration or the carriage are not depressing one or more of the switches enough to close the contacts.

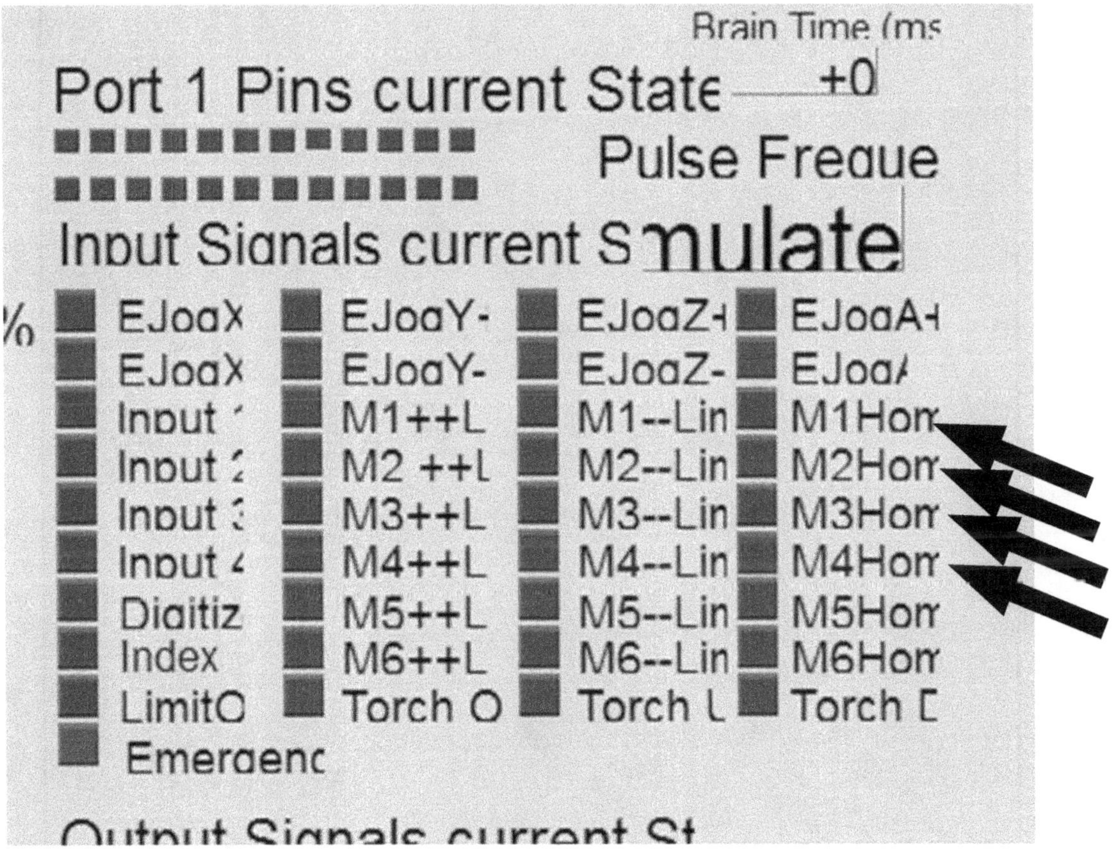

Figure 13.7

Conclusion

All your switches are now operational. Most likely the X and Z switches are adjusted properly as it does not take much to get them where you want them. This is not true of the Y and A switches. A little time is needed to get them adjusted properly. By moving the Y or A switch you can effectively square the X axis to the Y.

This can be checked by aligning the Y axis fence to the router, and then the X axis fence to the router. You then place a framing square against the two fences and if its not square you can move one of the two switches in or out slightly to make adjustments.

Just keep in mind that you need to rehome and adjust your fences with each test.

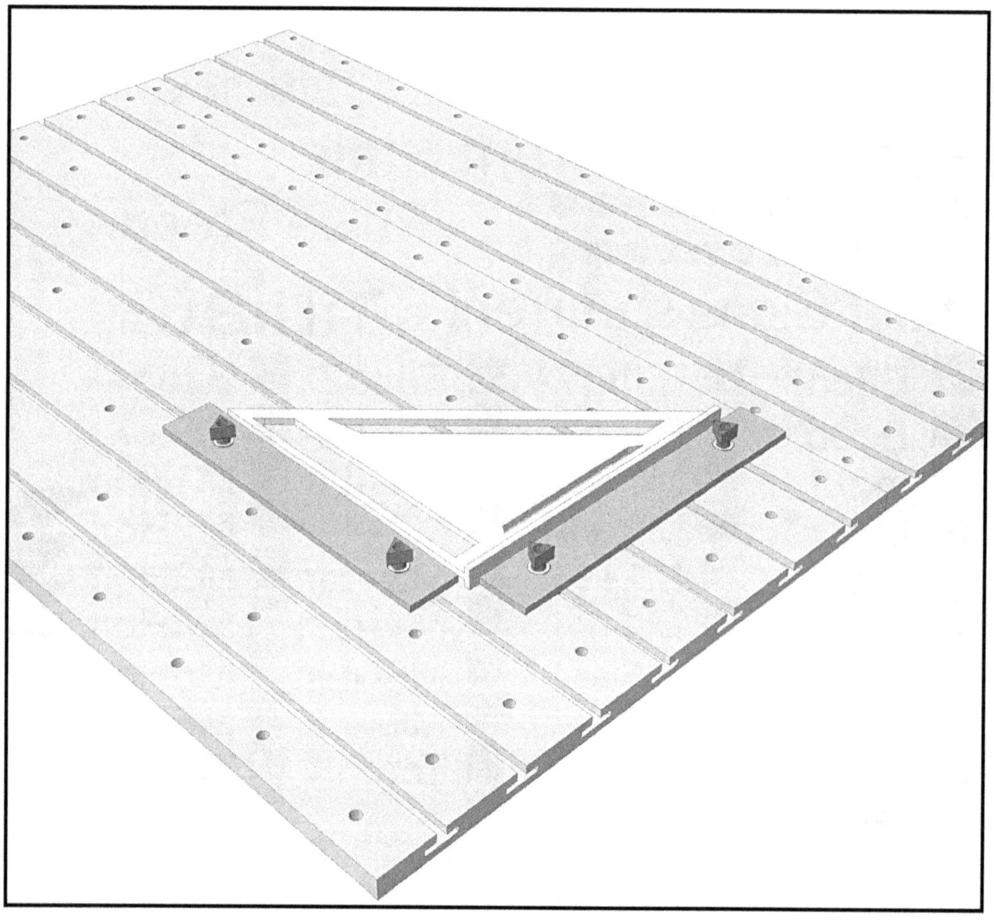

Figure 13.8

Chapter 14

Y-carriage Brace

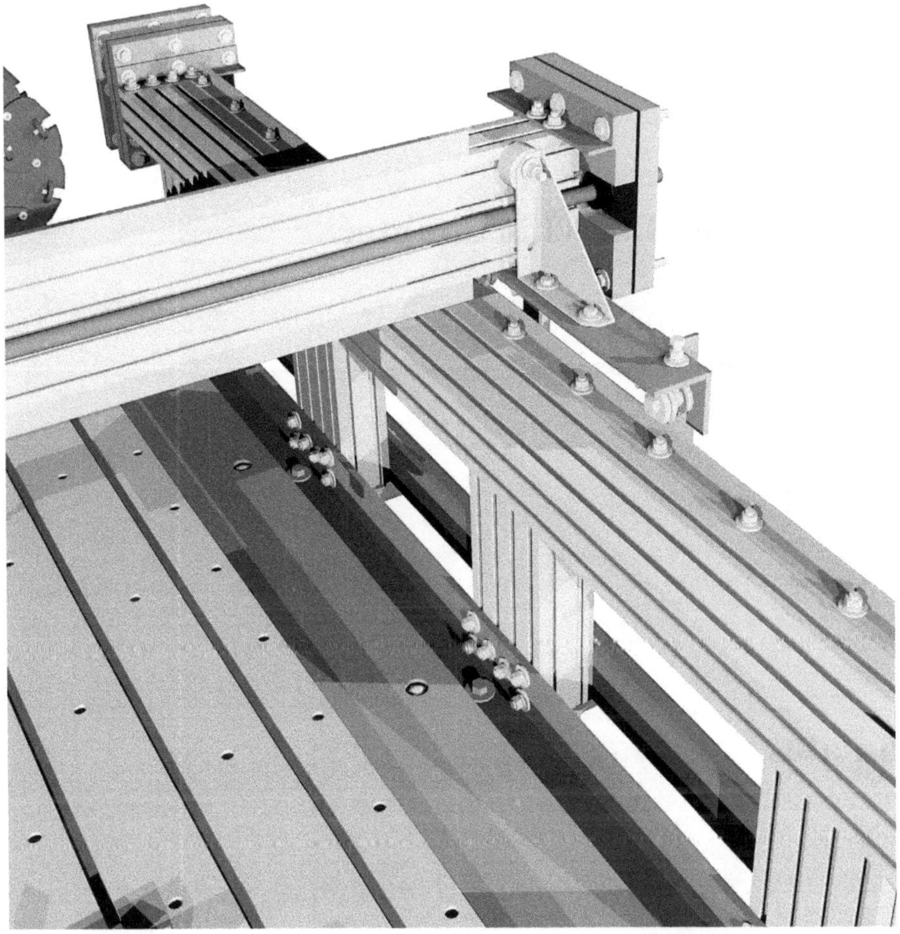

In this chapter you will add two braces to the Y-carriages on your KRMx01 CNC. This will stiffen the X beam and allow you to cut faster and more accurately on your CNC.

Tools Needed For This Chapter

- 2, 7/16 Wrenches or sockets
- Power drill
- 1/4" Drill bit
- 3/8" Drill bit

Components Needed For This Chapter

- 2, 4" Corner braces (Home Depot #831301)
- 2, 1/4-20 x 1-1/2" Carriage bolts
- 4, 1/4-20 x 1" Hex bolts
- 10, 1/4" Washers
- 10, 1/4" Split lock washers
- 10, 1/4-20 Hex nuts
- 20, 3/8" Washers

Time Estimates

Cost: $10

Time: couple hours

Prerequisites

It is important that you have a 1/4-20 x 1-1/2" carriage bolt installed into the front top T-slot on each side of the X-beam. You were asked to do this in Chapter 6 of the first KRMx01 book.

Step 1 - Prepare X-beam Brace Bolt

In the first book in the series you were told to add a 1/4"-20 x 1-1/2 bolt to the front on each side of the X-beam, as shown in Figure 14.1. If you did not add this bolt, you will have to remove the right and left bearing blocks on the X-beam, then add the bolt.

As an alternative you can drill a hole all the way through the extrusion. This will allow you to slip a 1/4-20 x 3-1/2" bolt through the extrusion.

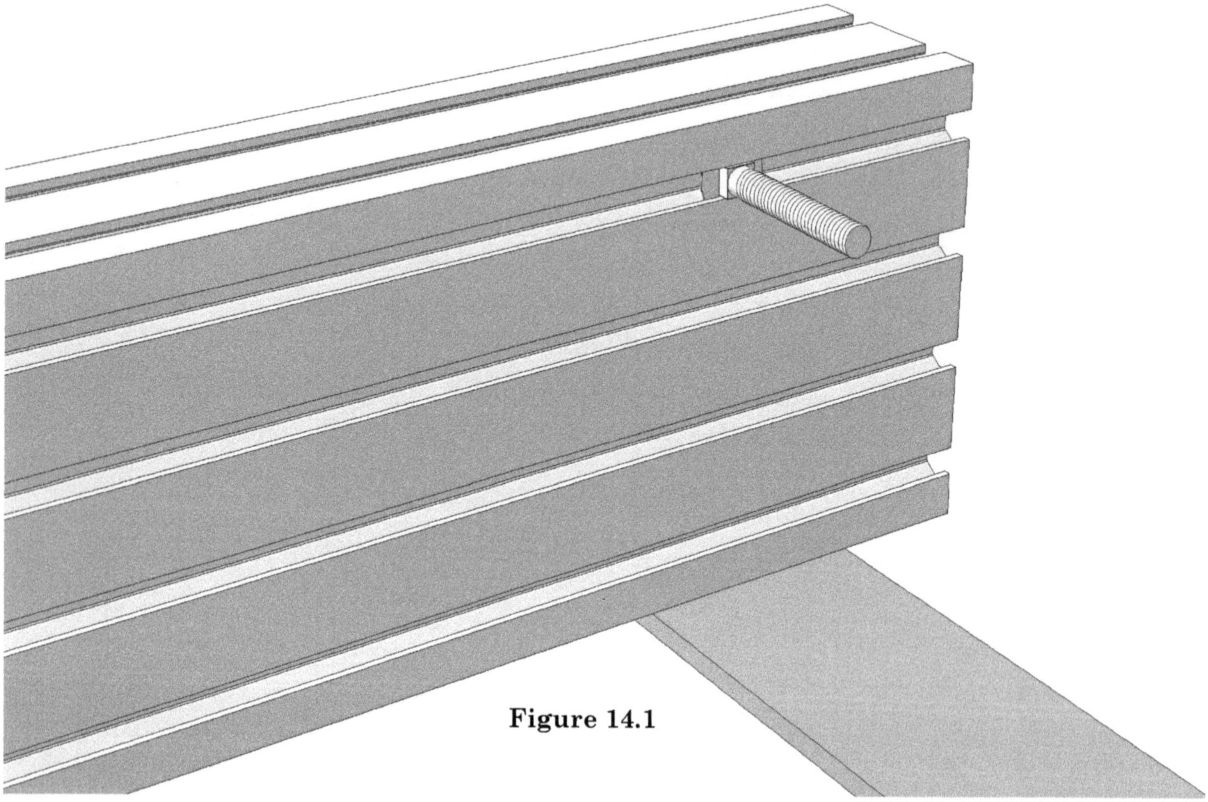

Figure 14.1

Step 2 - Position Brace

Slip 8, 3/8" washers over the carriage bolt, and slide one of the braces against the bolt, as shown in Figure 14.2. If the hole in the brace does not line up with the bolt, you will have to widen the hole with a 3/8" drill bit.

The washers are used to separate the brace from the extrusion in order for the leadscrew to fit. Add or remove washers so that you have about 1/4" clearance between the brace and the leadscrew.

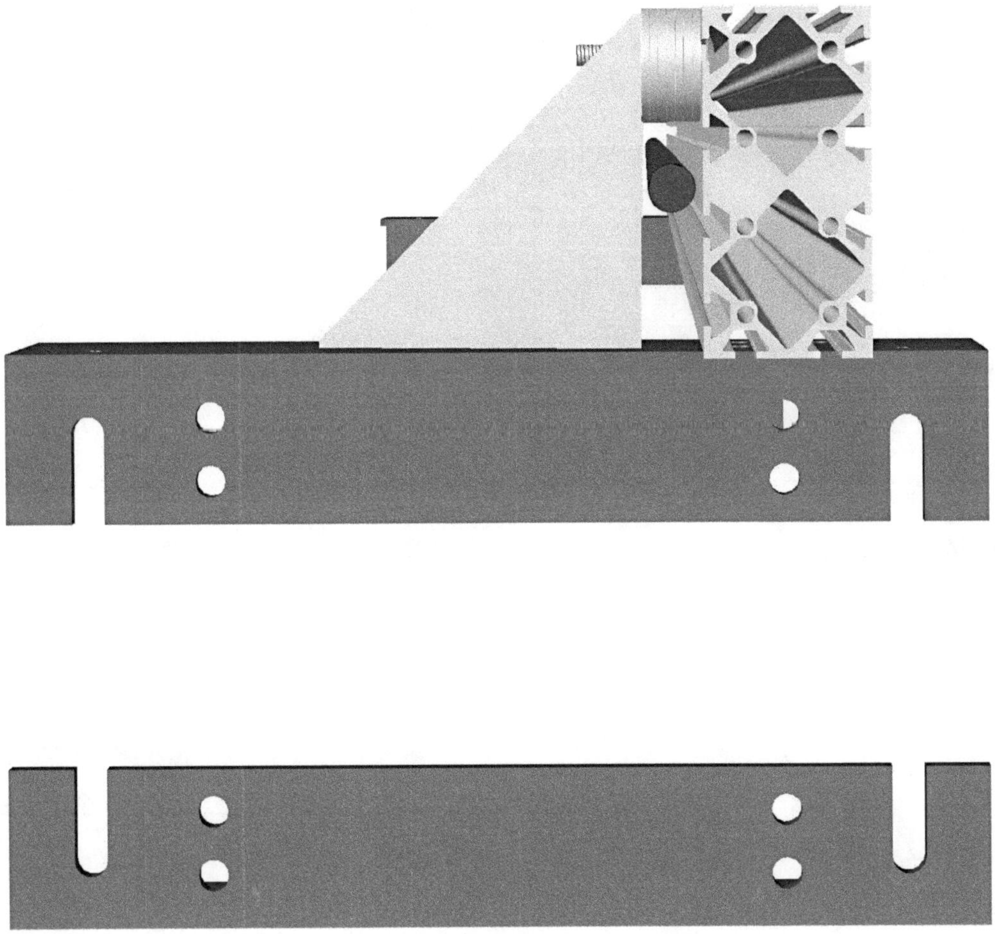

Figure 14.2

Step 3 - Add Hardware to Brace

Position the bolt and brace so that there is about 1/4" clearance of the brace to the inside edge of the carriage, as shown in Figure 14.3. Add a 1/4" washer, split lock washer, and hex nut to the brace as shown, and tighten.

Note that the bearing blocks and hardware have been removed from the illustration for clarity.

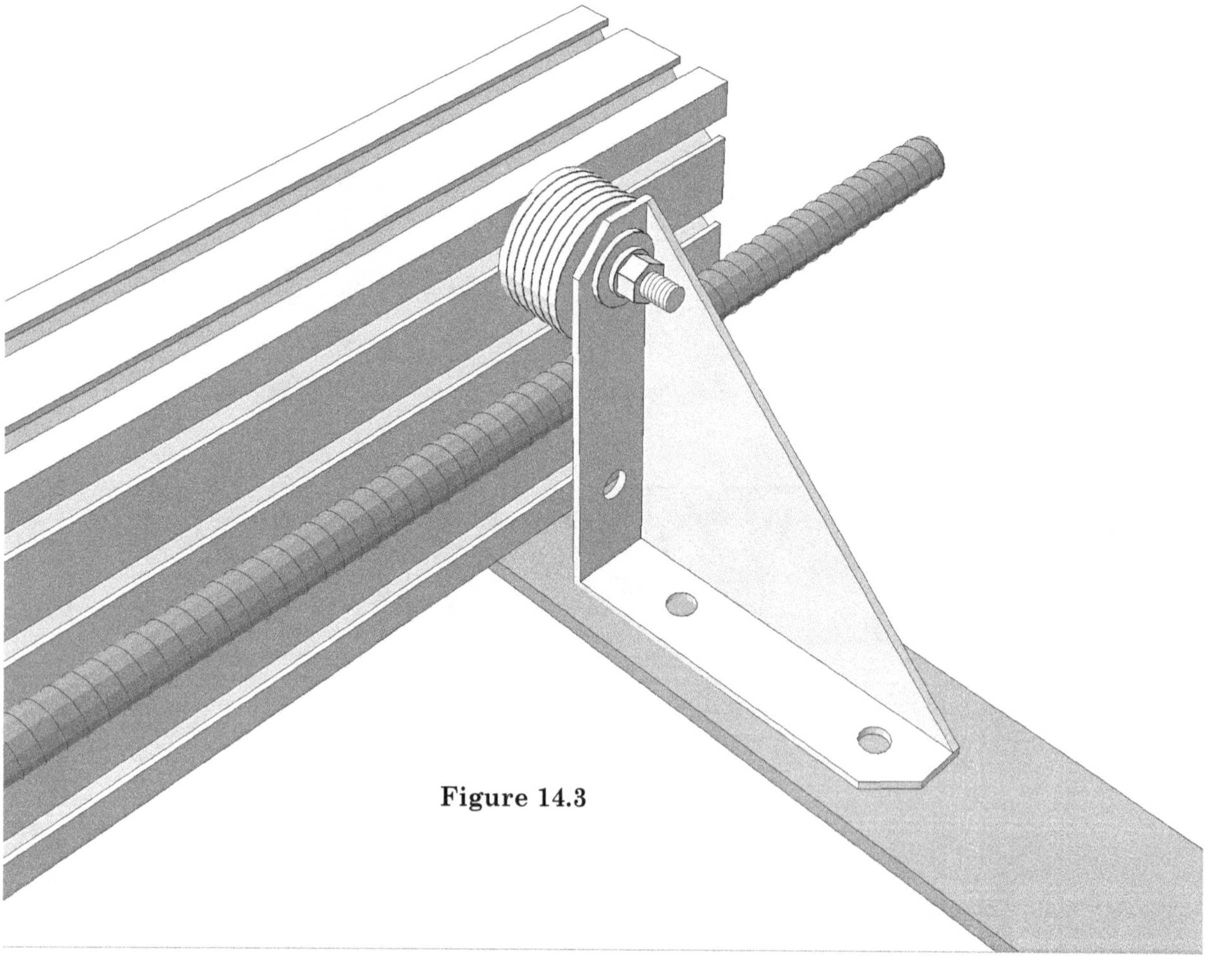

Figure 14.3

Step 4 - Drill Holes

Using a 1/4" drill bit in your power drill, drill two holes into the Y carriage, using the brace as a guide, as shown in Figure 14.4.

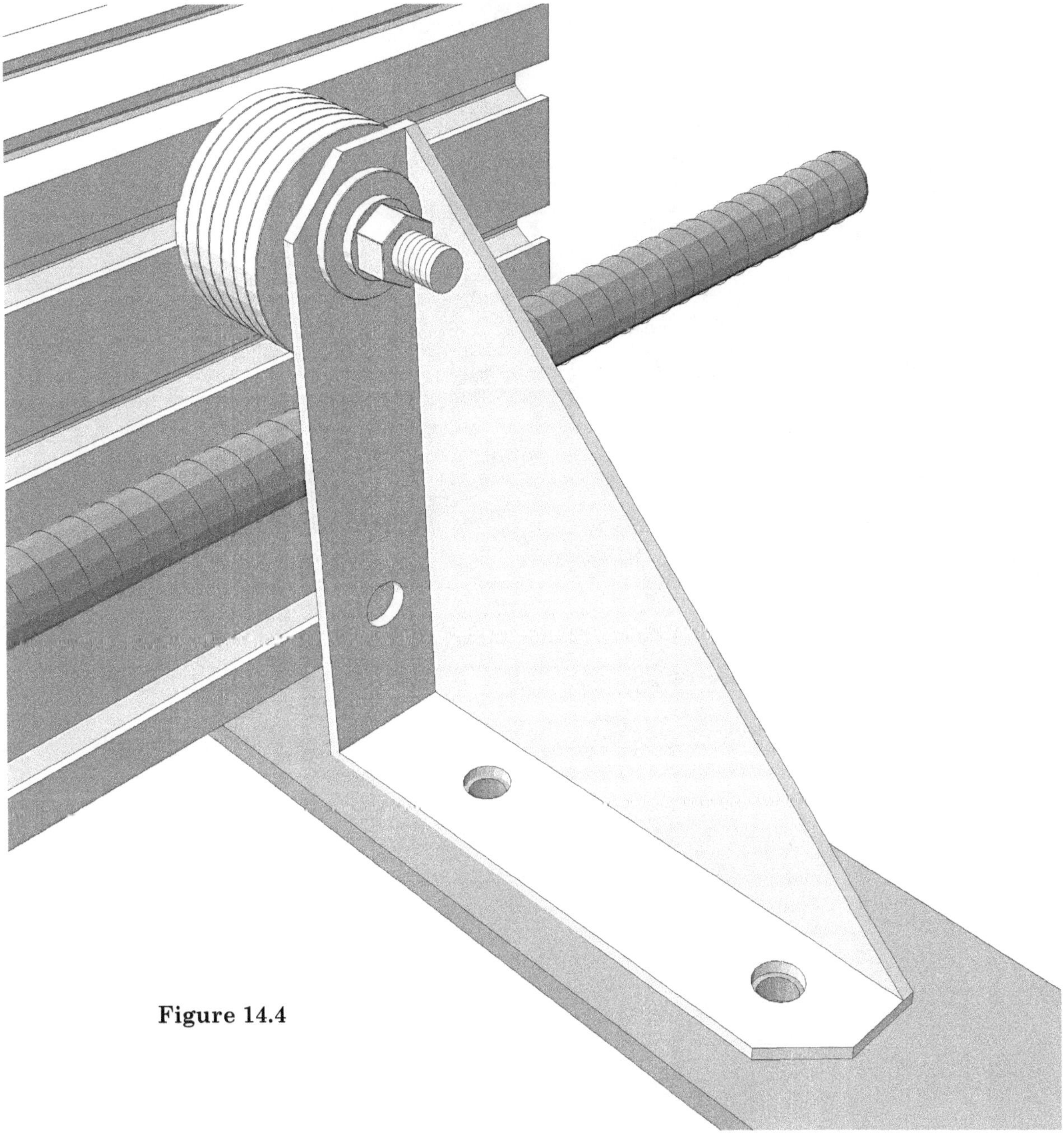

Figure 14.4

Step 5 - Add Bolts to Brace

Insert the 1/4-20 x 1" hex bolts with washers into the two newly drilled holes. Add a washer, split lock washer, and hex nut and tighten. It is important that the brace be flat against the Y carriage before you tighten the nuts. If not loosen the carriage bolt and press the brace down while retightening.

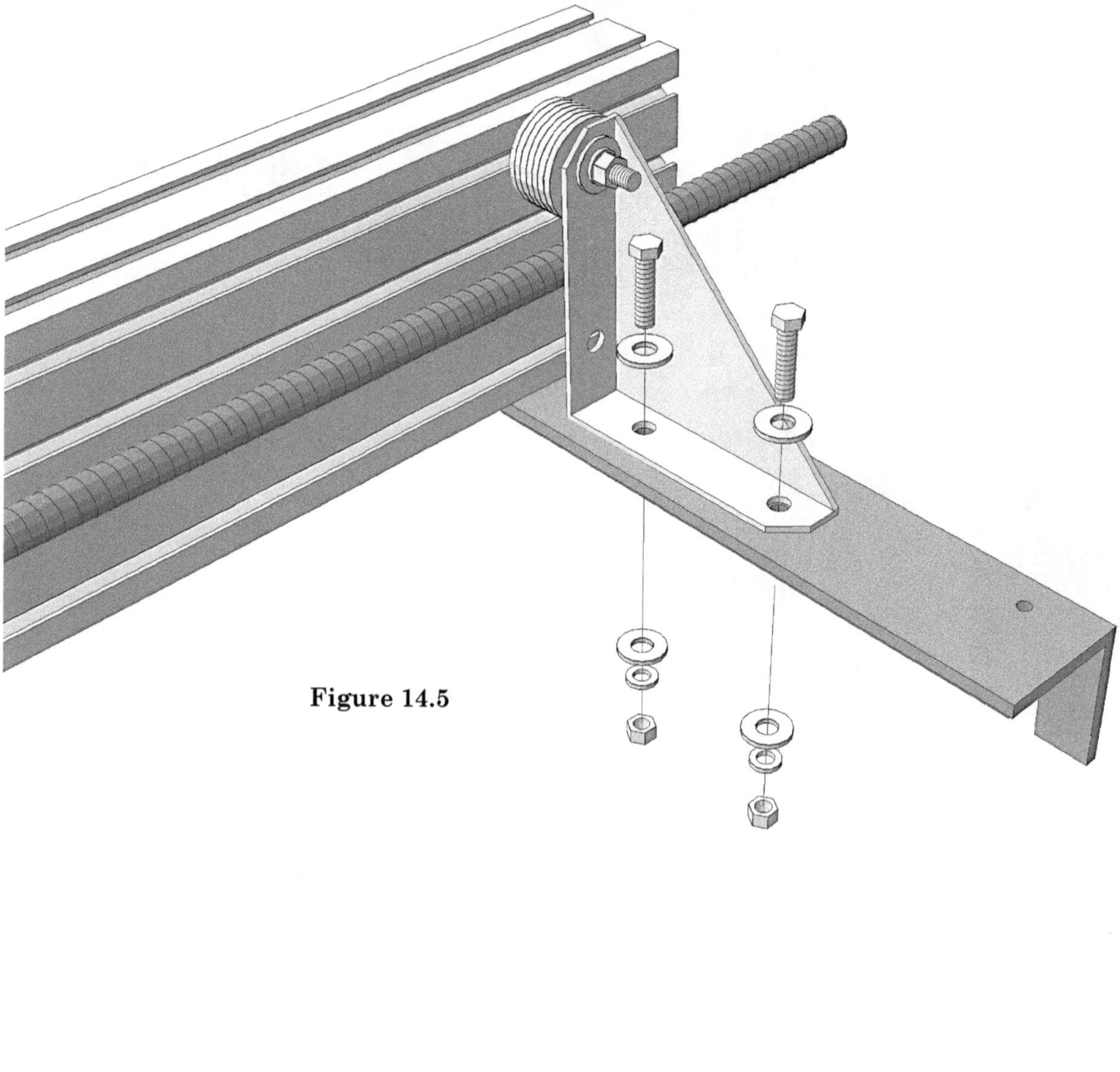

Figure 14.5

The completed brace should look like the one shown in Figure 14.6.

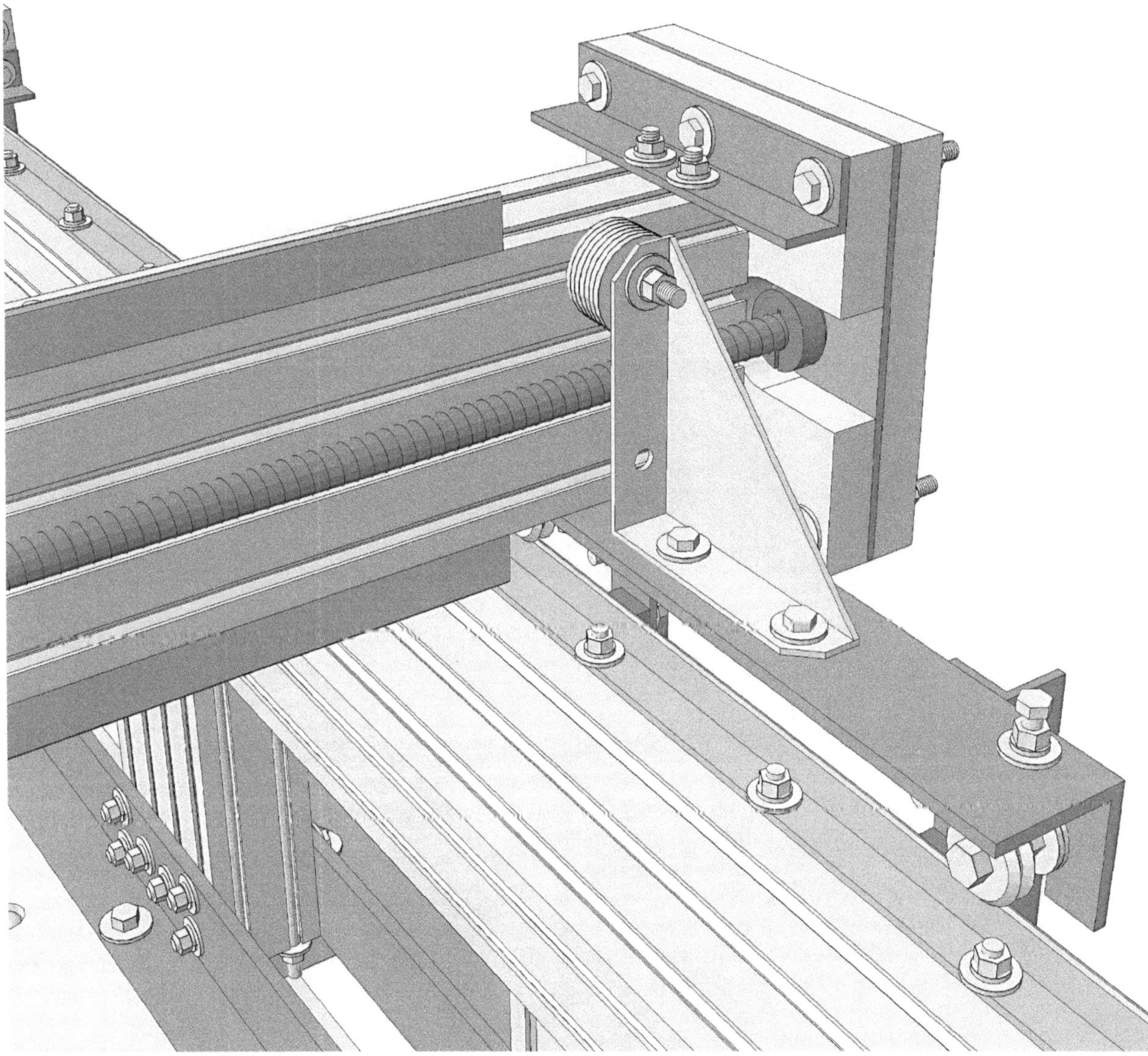

Figure 14.6

Conclusion

Repeat the process on the other Y-carriage.

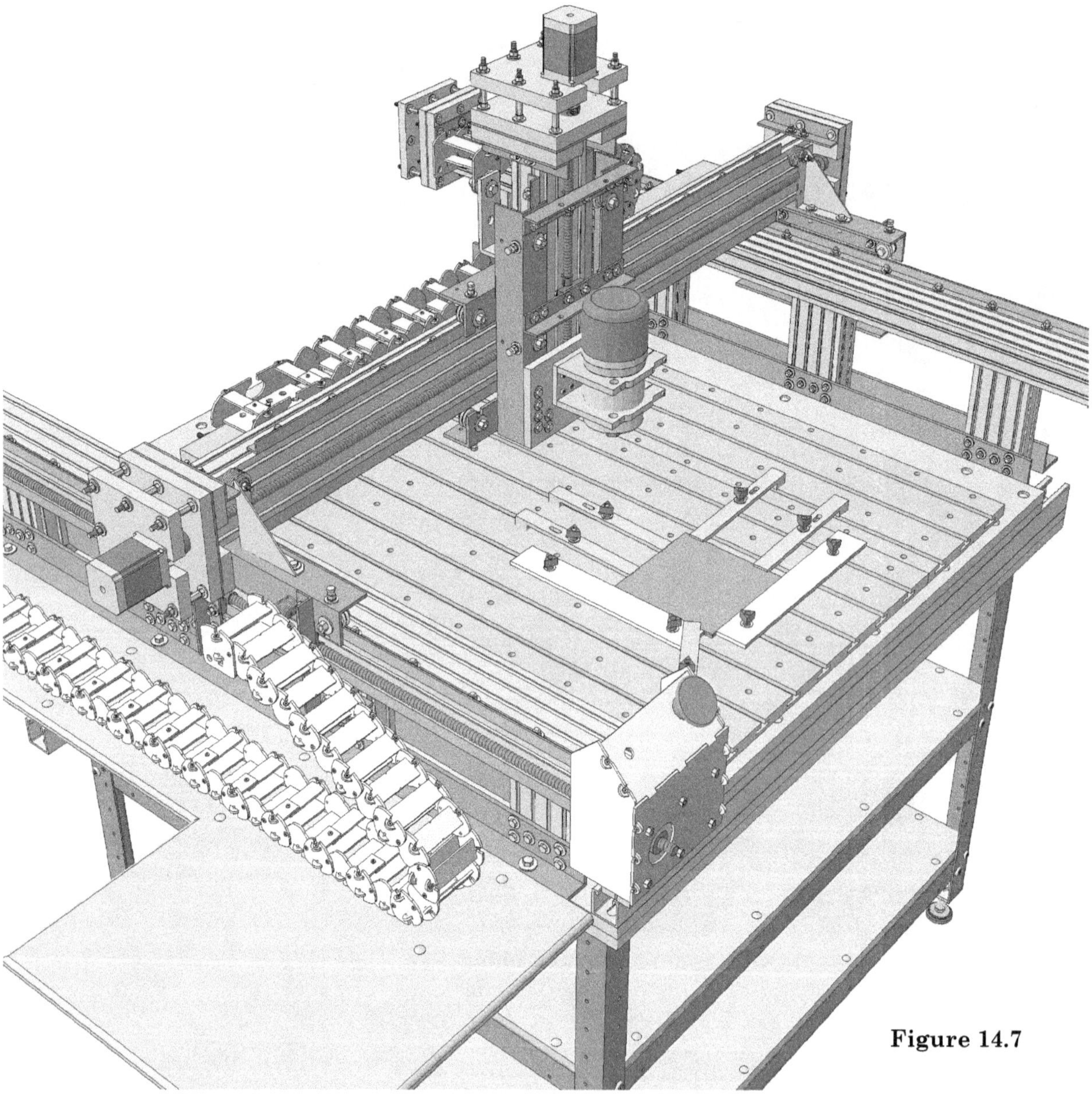

Figure 14.7

Chapter 15

Conclusion

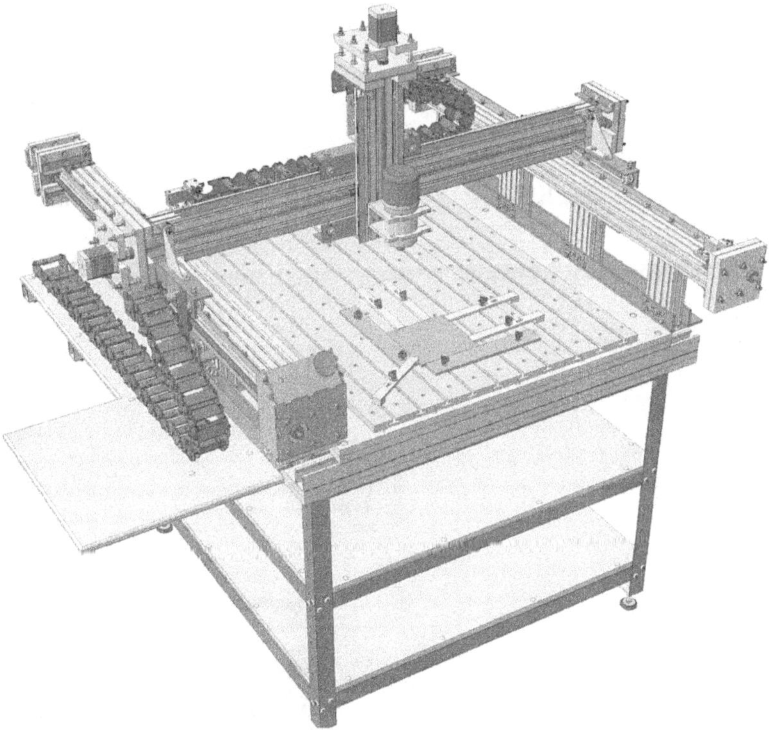

Congratulations! You have just taken your KRMx01 CNC to the next level.

In book three of the KRMx01 series you will be presented with even more upgrades. Probes, Xbox 360 Controller, and the AirExchanger dust collection system are just a few of the upgrades you will add.

KRMx01 Series

The KRMx01 series consists of the following three books:

- Building the KRMx01 CNC
- Upgrading and Operating the KRMx01 CNC
- More Upgrades for the KRMx01 CNC

For more information on the KRMx01 Series and other enhancements to your KRMx01 CNC, visit the Kronos Robotics web site at:

http://www.kronosrobotics.com

In addition to the KRMx01 design, Kronos Robotics also has a KRMx02 design. You can find more information on the KRMx02 here:

http://www.kronosrobotics.com/krmx02/

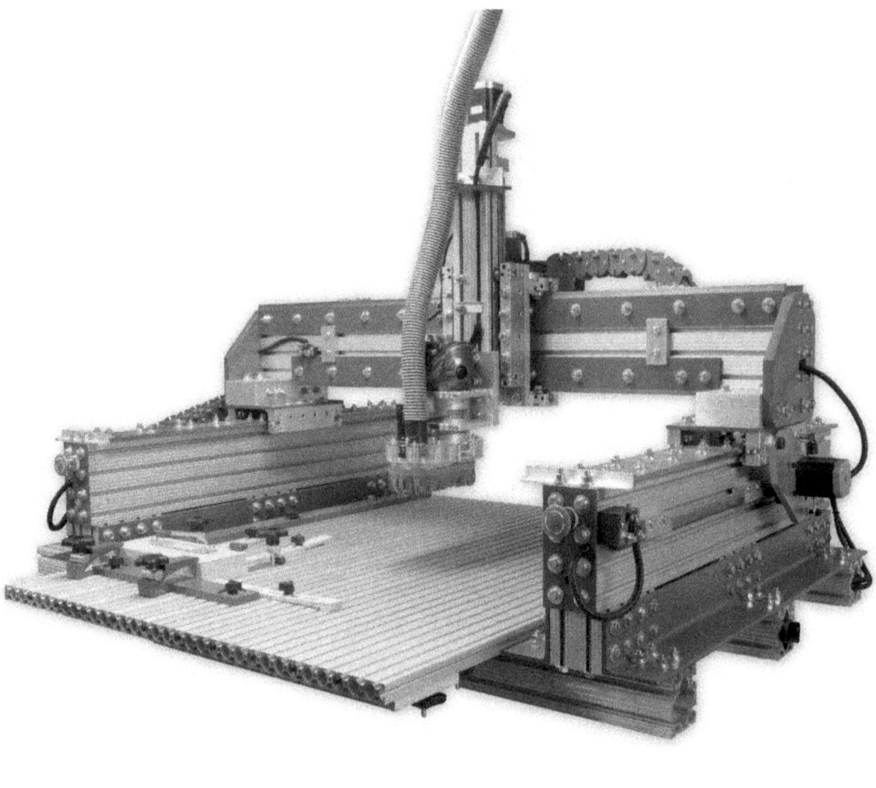

www.ingramcontent.com/pod-product-compliance
Lightning Source LLC
Chambersburg PA
CBHW081126180926
51611CB00004B/61
9781938687112